Kinematic Labs with Mobile Devices

Kinematic Labs with Mobile Devices

Jason M Kinser

Department of Physics and Computational Data Sciences,
George Mason University, Fairfax, VA, USA

Morgan & Claypool Publishers

Rights & Permissions
To obtain permission to re-use copyrighted material from Morgan & Claypool Publishers, please contact info@morganclaypool.com.

ISBN   978-1-6270-5628-1 (ebook)
ISBN   978-1-6270-5627-4 (print)
ISBN   978-1-6270-5630-4 (mobi)

DOI   10.1088/978-1-6270-5628-1

Version: 20150701

IOP Concise Physics
ISSN 2053-2571 (online)
ISSN 2054-7307 (print)

A Morgan & Claypool publication as part of IOP Concise Physics
Published by Morgan & Claypool Publishers, 40 Oak Drive, San Rafael, CA, 94903, USA

IOP Publishing, Temple Circus, Temple Way, Bristol BS1 6HG, UK

Dedicated to Sue Ellen

Contents

Preface

For decades an undergraduate education in physics has been accompanied by a laboratory experience that is essential in understanding the scientific process. Education, however, is undergoing major changes due to serious economic pressures and an onslaught of affordable and accessible technologies. One response to these pressures is to offer education on-line which is plausible for lectures but more difficult for labs.

The in-lab experience is still a superior method of providing the proper experience but modern times will insist that some students obtain the lab experience through some other method. Creating such labs suitable for off-site implementation will require that the equipment the students use be readily available and inexpensive. This includes the items used in the experiment as well as the sensors.

Quick surveys in freshmen level physics classes at George Mason University indicate that about 75% of the student population has a smart device (phone or tablet) and that almost all students have personal computers. These smart phones have a multitude of sensors that are suitable for kinematic labs. So, within the economic pressures pushing education on-line and off-campus along with the plethora of sensors on personal devices this text finds its justification.

The intent of this text is to lay a foundation for freshman level labs that can be performed at home for on-line students. These labs are based on using a personal smart phone to collect the data from experiments easily created outside the lab environment. The necessary equipment can be purchased at large discount stores or hobby shops. Most of the items cost only a few dollars with the most expensive items being a basketball and two toy train cars. The combined cost for all labs can be less than $100 excepting, of course, the computer and phone.

This text is not intended to replace in-lab experiences, but instead is designed to be a guide for those situations where an in-lab experience is not feasible. Instructors should feel free to modify the labs and I will be delighted to hear of the successes achieved with these labs.

Jason M Kinser
George Mason University
jkinser@gmu.edu

Acknowledgements

The author greatly appreciates the input from Sean Francis, Suzanne Hewitt, Connor Maloy, Sejal Naik, Nicholas Pellegrino, Benjamin Quann, Sergio Ribeiro, James Stowell, Justin Williams and Andy Zhu.

The author also wishes to thank Dr Michael Summers for his encouragement at the beginning of this project.

Author biography

Jason M Kinser

Jason M Kinser, DSc, has been an associate professor at George Mason University (GMU) for over eighteen years teaching courses in physics, computational science, bioinformatics and forensic science. Recently, he converted the traditional university physics course into an active learning technology environment at GMU. His research interests include modern teaching techniques, more effective methods in text based education, image operators and analysis, pulse image processing and multi-domain data analysis. This book was born from a desire to engage students in physics education and to find ways of reducing the external costs that both students and institutions incur within the traditional education framework.

Part I

Lab preparation

Chapter 1

Apps for mobile devices

Mobile devices such as smart phones are quite prevalent in our modern society. These devices come equipped with many sensors that will be useful in collecting data for physics labs in mechanics. While the sensors are already in the devices the apps necessary to store the data are generally absent. Thus, it will be necessary to download the appropriate apps. It would be fruitless to list the apps that are useful since the apps are evolving at a rapid pace. The list would be outdated before this text reaches publication. Therefore, this chapter will focus more on the requirements of the apps.

1.1 Operating systems

Currently, there are two main operating systems for smart phones: Android and Apple iOS. Other operating systems are available, but these are not yet prevalent in the student population. There is not a significant difference in performance between apps running on these two operating systems. Cost-free apps do seem to be more abundant for Android systems, but the cost of apps for iOS systems is only a few dollars.

1.2 Typical sensors

Smart phones contain a sensor suite that is used in a variety of ways for different phone apps. For example, when a user turns a phone sideways it rotates the display, which is a programmed response when a sensor detects a major change in the phone's orientation. There are several sensors which will be used here to collect data in physics experiments. There are, of course, the overt sensors such as the camera and the microphone, but there are also many sensors that may not be so obvious to the casual user. For example, there are accelerometers, orientation sensors, magnetic field sensors and so forth. The sensors that are used in the labs in the following chapters are reviewed in the following subsections.

doi:10.1088/978-1-6270-5628-1ch1   1-3

1.2.1 The camera

Smart phones have at least one camera that can be used in three different modes. Newer phones may also have a forward facing camera which usually has a poorer performance than the camera that is placed on the back of the phone. There are three modes or types of uses for the camera. The first is the camera mode which allows it to take high resolution single images. This mode will be used to collect photographs of the experiments which are then analyzed by software (see section 1.3.3). The second mode is the video camera mode which collects data at a rate of about 30 Hz. Usually, the video frames are of poorer resolution than the still photographs. These are used to collect data during an experiment that involves motion of an object. The third mode is the *burst* camera mode which requires an additional app. The burst camera captures still frames at a high rate, usually between 12 and 20 Hz. These frames are of better quality than those of the video mode and are saved with a time stamp. There are multiple apps that can retrieve burst images and they usually have the word 'burst' in their name.

1.2.2 Audio

Each phone has a microphone (which is required in order to act as a phone) that can be used to capture the sounds that occur during an experiment. An example would be to capture the sounds of a ball bouncing on the floor and then using the audio track to determine the times at which the ball hit the floor. The advantage of the audio capture is that it has a much higher sampling frequency and thus can place the time of an event more accurately than the camera. None of the labs in this text specify the use of the audio device but it could certainly be used instead of the recommended sensor if the instructor so desires.

1.2.3 Orientation sensor

The orientation sensor relays information about the orientation angle of the phone. This is a three-axis sensor and so there is information about roll, pitch and yaw as shown in figure 1.1. It should be noted that some apps only show the angle to the nearest degree which is one cause of uncertainty in measurements.

Some phones have a protrusion on the back surrounding the camera lens. In this case the phone will not lie flat and therefore the orientation sensor might read a degree or two while the phone is in this position. These phones have a small bias which needs to be removed during calculations.

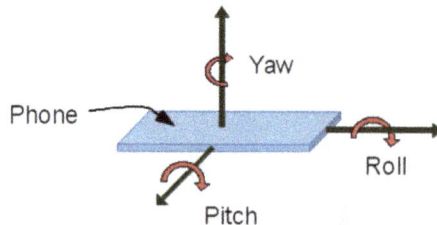

Figure 1.1. The phone has three orientation axes and the rotations about each are the roll, pitch and yaw.

1.2.4 Accelerometer

The accelerometer measures the acceleration along three axes. Many apps will include acceleration due to gravity in their displays thus the three accelerations from a phone lying flat will be about $0 \, \text{m s}^{-2}$, $0 \, \text{m s}^{-2}$ and $9.8 \, \text{m s}^{-2}$. These sensors are not very accurate and there may be a bias of $\pm 0.3 \, \text{m s}^{-2}$ which will need to be considered during the calculations. Furthermore, the sensors do not react very fast and so there is an upper limit on the change in acceleration that can be detected.

When using the acceleration it is important to align the phone such that one of the axes is aligned with the motion. If the phone is at an angle to the motion then the acceleration will appear in two axes of the collected data. While the data are still collected it will require extra effort to extract the needed values.

1.2.5 File manager

A file manager is a program that will display all of the user files through a directory structure much like a PC. This app is useful because it has the ability to *share* files with other devices. It is through the file manager that data are sent to the user's PC for analysis. Usually, the data from an app are stored in a directory with the same name as the app. The user presses on the file and holds until a menu pops up with options including the ability to email the file or upload it to a cloud.

1.2.6 App requirements

Apps are constantly evolving as upgrades or new replacement products and so it is not possible to indicate which apps are the best for physics experiments. However, it is possible to relay the requirements of the apps as used in these experiments.

Orientation apps display the angle of the phone relative to the horizontal in all three axes. There are two detection functions that are required from an orientation app. The first is to simply display the orientation on the screen. In some labs the app is used by aligning the phone with an object in the experiment to obtain the orientation of the object. The second function is to log orientation angles while the object is moving. In this case the app will need to log the readings into a file which can later be sent to a computer. It may, therefore, be necessary to download two different apps.

The accelerometer, on the other hand, will be tasked to collect data at a high sampling rate. It will be necessary for these data to be stored in a file for later analysis. Several apps that display accelerometer data do not store the information in a file and so these are not sufficient for these experiments. Usually, the name for a useful app will contain the word 'logger'.

Another issue with apps that capture accelerometer data is that they may also be collecting data from other sensors. Thus, the data stored in a file may have acceleration interlaced with data from the magnetic sensor and the orientation sensor. To further confound this issue, the sensors are sampling at different rates and thus there is no guarantee that the placement of the acceleration data in the file follows a rigid pattern. These sensors will create a text file that shows the detected values from the sensors interlaced with each other. If the app is collecting data from all sensors then it needs the ability to select only the sensors of interest.

1.3 Analysis software

Finally, it will be necessary to have software that will perform the analysis of the data. This software must reside on a computer as such software is not available for phones. Even if such apps were available it would be very cumbersome to perform any meaningful analysis on a cell phone. A few programs are needed with most being free of cost. The requirements of the programs will be discussed in the following and some products will be named.

1.3.1 Data analysis

The only software that is required for analysis is a professional spreadsheet (such as Microsoft Excel or LibreOffice Base). Of the two, Excel has more tools for analysis but both will work for these experiments. The spreadsheets can manipulate the data, perform calculations, create plots and graphs, and even estimate the function that best describes plotted data.

1.3.2 Image extractions

In the cases where the data are received by a video camera it may be necessary to extract some of the image frames and the time stamps of the frames. Programs such as Microsoft MovieMaker or Sony Vegas can accomplish this task. While some programs can extract a frame from a movie they do not necessarily indicate at which time during the video the frames were extracted. These products will be insufficient for use here. An alternative to video capture is to use the burst camera mode which requires a separate app (see section 1.2.1).

1.3.3 Image measurements

It will be necessary to gather the height of objects in a frame. This is accomplished by including a ruler or meter stick in the frame along with the experiment. However, the next step is to correlate the length of the ruler with the length of the objects in the frame. This is accomplished by comparing the number of pixels that each object extends. Therefore, it is necessary to obtain image viewing software that at least displays the position of the mouse in the frame. Free programs such as PhotoFiltre have this option. A better choice is GIMP which provides a tool named 'Measure' (Shift-M) which allows the user to drag the mouse from one location to another and displays the distance in pixels and the angle to the horizontal.

There are some issues with measuring objects in an image since the focal plane of the camera is actually a curved surface rather than a plane. Consider figure 1.2 which shows the camera on the left and a long rectangle that is to be measured on the right. The arrow near the bottom is the length of a meter stick and the viewing angle is θ. The viewing angle is directly related to the number of pixels that the length consumes in the camera. At the top is another arrow of exactly the same length with the viewing angle ϕ. It is seen that $\theta > \phi$ which indicates that the number of pixels consumed by each arrow is different even though the vertical lengths are the same. So, the method of measuring objects using pixel lengths is only accurate for lengths comparable to the ruler length. When taking an image to

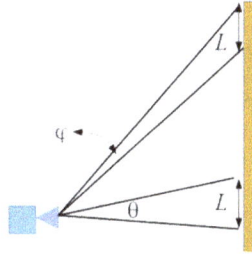

Figure 1.2. The lengths seen by the camera are different depending on the viewing angle even though the lengths of the arrows are the same.

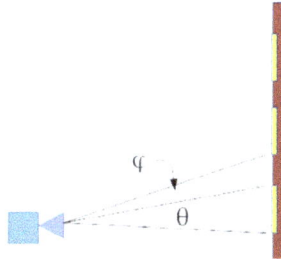

Figure 1.3. Measuring the height of a building can be performed by obtaining the height of the lowest window and the distance between windows.

gather length information it is important to position the camera perpendicular to that length.

There are other options, however, for measuring large heights such as the height of a building as shown in figure 1.3. In this case the student can measure the height of a single window and the distance between them using the method above. The upper windows will extend over fewer pixels in the camera but it is known that they are the same height as the lowest window. The same is true for the distances between windows. Thus, by measuring the first window the student has all of the height except the portion below the lowest window and the portion above the highest window.

1.4 Example

This simple example demonstrations the typical appearance of the raw data and the steps necessary to convert these into usable data. Again apps will differ in the format of the data but this example shows a common presentation. In this example, a student has the very easy task of using the accelerometer to collect measurements of gravity. The app (in this case Accelogger) collects the data and the file manager is used to share the data with the students. These data are then opened in a spreadsheet and a portion of them are shown in figure 1.4. This particular app has five columns. The first two columns are the time stamp in two different formats with the second column showing the time in nanoseconds. The last three columns are the measurements of acceleration in three axes.

▲	A	B	C	D	E	F
3	10/19/2014 18:28	2.56608E+11	-0.38137	2.329079	9.302697	
4	10/19/2014 18:28	2.56622E+11	-0.217926	2.369941	9.343558	
5	10/19/2014 18:28	2.56638E+11	-0.108963	2.410802	9.343558	
6	10/19/2014 18:28	2.56656E+11	0.027241	2.478903	9.41166	
7	10/19/2014 18:28	2.56671E+11	0.149824	2.560625	9.575105	
8	10/19/2014 18:28	2.56685E+11	0.190685	2.601486	9.643207	
9	10/19/2014 18:28	2.567E+11	0.190685	2.519764	9.724928	
10	10/19/2014 18:28	2.56714E+11	0.217926	2.478903	9.724928	
11	10/19/2014 18:28	2.56729E+11	0.299648	2.519764	9.833891	
12	10/19/2014 18:28	2.56744E+11	0.449471	2.601486	9.874752	
13	10/19/2014 18:28	2.56758E+11	0.640156	2.669588	9.915613	
14	10/19/2014 18:28	2.56772E+11	0.8036	2.75131	9.915613	
15	10/19/2014 18:28	2.56787E+11	0.8036	2.669588	10.02458	
16	10/19/2014 18:28	2.56802E+11	0.762739	2.519764	10.14716	
17	10/19/2014 18:28	2.56817E+11	0.681017	2.478903	10.29698	
18	10/19/2014 18:28	2.56831E+11	0.721878	2.519764	10.29698	

Figure 1.4. Typical raw data from an accelerometer. The first two columns show the times that the data were collected in two different formats and the last three columns show the detected acceleration for the three axes.

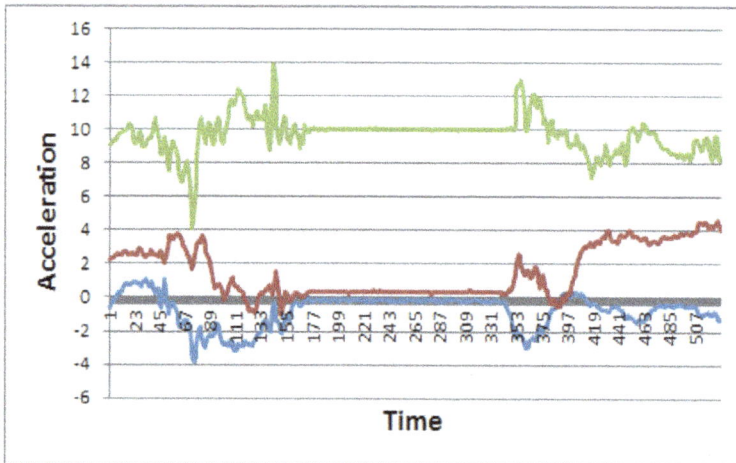

Figure 1.5. Plots of the raw data from the last three columns of figure 1.4.

The data are shown as plots in figure 1.5 which shows the acceleration along the three axes. However, there is also quite of bit of junk in this reading. In this case, the student turned on the sensor and then sat it down on the table. When the short experiment was over the student picked up the phone and stopped the recording. This action of moving the phone accounts for the huge variations at the beginning and ending of this data sample. The data need to be trimmed.

The student will need to identify which column contains the pertinent data. In this case the last column shows the data along the axis which is important to this experiment and the pertinent data from the experiment ranged from $x = 185$ to $x = 330$. Figure 1.6 shows the isolated data of this simple experiment which show

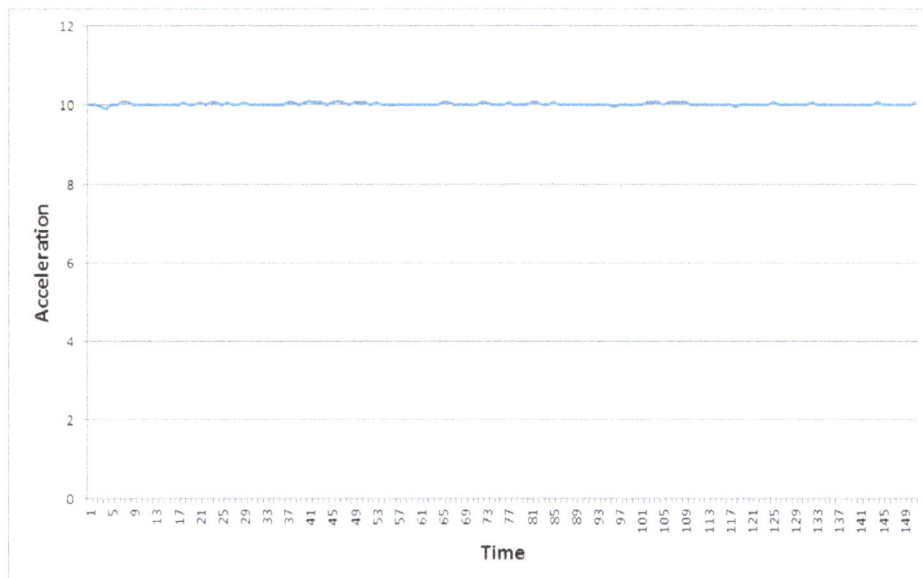

Figure 1.6. Data from the experiment.

a nearly constant reading of a value close to 9.8 m s^{-2}. Of course, this is a very simple experiment but it does show how the data from the smart phone must be pruned before any calculations can be performed.

1.5 Other materials

These labs are designed to be performed outside the standard lab setting. Most can be undertaken at home and a few can be performed at other locations as the students deem fit. For example, one lab uses an elevator which many people do not have at their residence. However, elevators are not that difficult to find or use.

The equipment used in these labs can be purchased at a discount store or a hobby shop. In some cases the students may have to be creative. For example, several experiments use weights and lab weights may be difficult to find. Creative solutions would be to use household items (a bag with pennies) as the weights. The mass of these can be measured from a scale or computed by looking up the weights of pennies on the Internet. In this case it should be noted that the weight of pennies changed in some years.

A set of scales suitable for measuring a few grams is also not a common household item. The best choice is to purchase a jeweler's scale of which there are dozens available through trusted websites. Electronic scales can also be found at sporting goods stores, particularly in the fishing section. However, many of these don't have precision down to 1 g which would render them useless for these labs.

The labs shown in the chapters are just one way of performing the experiments. Students should be allowed the freedom to adjust the labs to suit their situations and supplies. Such freedom in itself is a lab experience.

1.6 Summary

Smart phones have a useful suite of sensors but it will be necessary to download apps that can display or log the data they collect. These apps are evolving rapidly and so only the requirements for the apps are presented. Logged data will be stored on the phone as an obvious file which can be found using a file manager. This manager will have the ability to send the data file to a computer via a variety of sharing avenues (email, blue tooth, social pages, etc).

Chapter 2

Working with spreadsheets

While smart phones are excellent in many applications and are good tools for collecting data, they do not make good tools for developing analysis. Thus, the scheme is to transfer the collected data to a computer where a spreadsheet can be employed to perform the necessary calculations. The two best choices are Microsoft Excel or LibreOffice Calc. While there are several spreadsheets that can perform computations and create sufficient graphs, these two packages also provide curve fitting tools which are required for several labs. Spreadsheets such as Google Sheets do not currently offer these tools.

This chapter walks through the basics of performing calculations with Excel that are necessary for the remaining chapters. LibreOffice Calc offers the same functionality and very similar commands.

2.1 A cell

Cells in Excel are defined by column and row identifiers. The cell marked with a box in figure 2.1 is A1. The column and row identifiers are highlighted in yellow and A1 appears in the white box above the cells. The name of a cell can be changed by replacing the contents of this white box with a chosen name.

2.2 Simple math

It is possible to use the values in one cell in a computation for another cell. The following example is shown in figure 2.2. The equation starts with the equals sign and the cell that is being used is called by its name. In this case the contents of B1 will become the contents of A1 plus 5. The equation is also duplicated in the window at the top of the image. When the Enter button is pressed the computation will be completed. If the contents of A1 are changed then the contents of B1 will automatically be changed.

Consider the case of figure 2.3 in which column A has many values that were typed in and column B has the same equation from figure 2.2. The goal is to copy

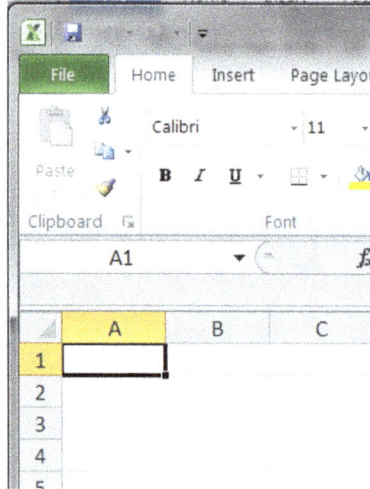

Figure 2.1. A single cell at A1.

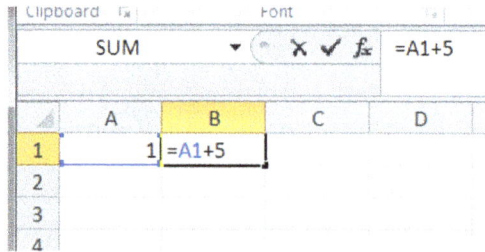

Figure 2.2. Addition. The contents of cell A1 are used to compute the value in B2.

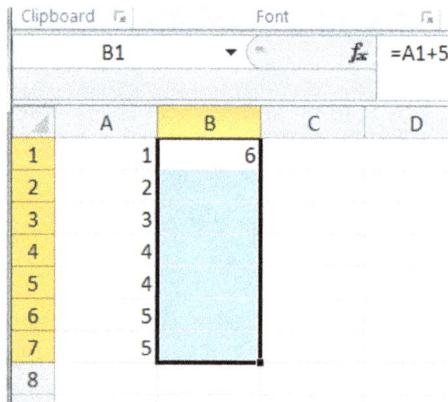

Figure 2.3. Copying the formula to many cells is performed by first painting all of the cells involved.

the equation in B1 to all of the other cells in column B but to also have them change their input cells. In other words, B2 is to be A2 + 5 and B3 is to be A3 + 5. This is accomplished by highlighting all of the cells that are to be changed and including the top cell which contains the formula as shown in figure 2.3. Then select the option to

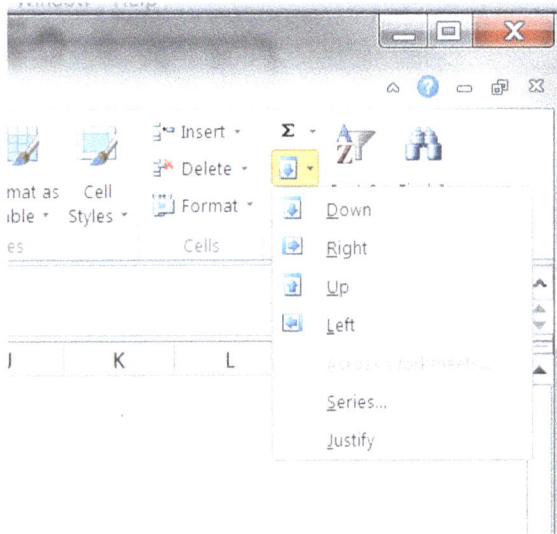

Figure 2.4. The second step is to select the fill down option from the menu.

Figure 2.5. The fill down results show that the computation is copied to all cells.

fill down as shown in figure 2.4. The shortcut for filling down is Ctrl-D, and the shortcut for filling to the right is Ctrl-R. The results are shown in figure 2.5. One of the cells is highlighted and the equation is shown at the top. Note that the equation for B4 uses cell A4.

In some cases it may be necessary to not have each equation in the fill down area change the cells that are being used. If, for example, all equations are to use cell A1 instead of changing, then the dollar sign is employed. An example is shown in figure 2.6(*a*) in which dollar signs are placed in front of the A and 1 so that when the equations are copied in the fill down the cell A1 is used in all of the cells. This is shown in figure 2.6(*b*).

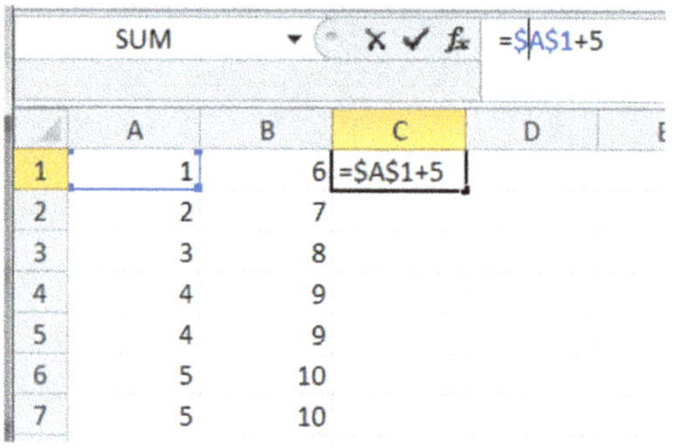

(a)

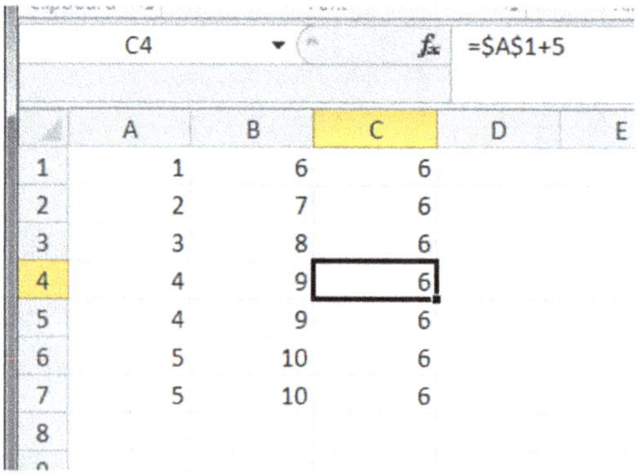

(b)

Figure 2.6. Keeping the same cell identifier requires the use of the dollar sign.

In this equation both the A and the 1 are constant. It is possible to put the dollar sign in front of just the A so that the column identifier remains constant. If the dollar sign is in front of just the number 1 then the row identifier remains constant.

2.3 Plotting

Spreadsheets have many tools for creating informative graphs and plots. One such tool used here is Scatter which the user starts by painting two data columns as shown in figure 2.7. In a scatter plot the first column will correspond to the x-axis and the second column to the y-axis. One mistake that is easy to make is to have the columns in an incorrect order thus creating an inverted plot, so it is important to place the columns of data in the correct order.

Figure 2.8 shows the result which depicts, in this case, a linear plot. The option of showing the tick marks with the line was chosen whereas the example shown in figure 2.7 would have created a plot without the markers on each data point.

The graph has several areas that can be adjusted. For example, a right mouse click (for PC users) on the x-axis will allow the user access to modify this axis. These modifications include the range, number of tick marks, fonts, etc. Other hot areas include the labels, colors, appearance, axis lines and other aspects

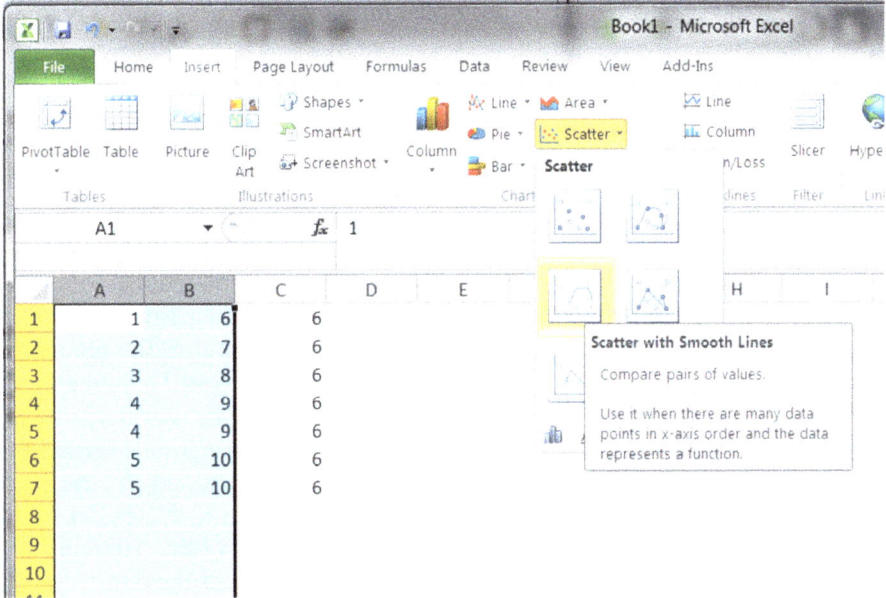

Figure 2.7. Creating a scatter plot graph in which the data in column A will correspond to the *x*-axis and the data in column B will correspond to the *y*-axis.

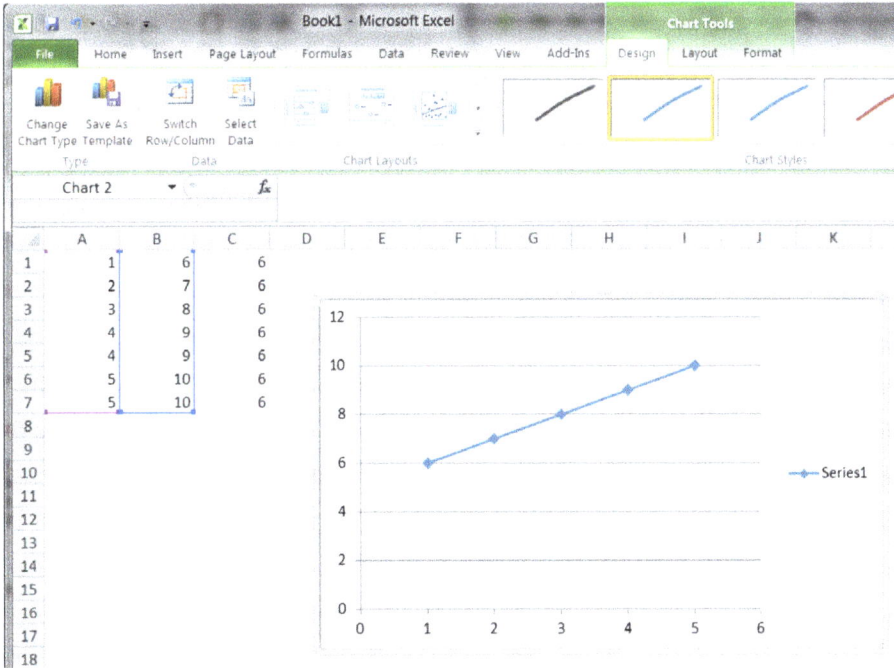

Figure 2.8. The graph resulting from figure 2.7.

of the graph. Once perfected it can be copied into a word processor or saved as an image.

This is a just a small example of the tools that these spreadsheets can offer. However, for the labs in this text simple plotting is sufficient and so a more detailed look into the power of a spreadsheet is not warranted. There are, however, two tools used for estimating functions that need to be discussed as they are useful in many of the labs.

2.4 Trend lines

In a few experiments it will be necessary to estimate the function that best fits the raw data. The Trendline tool provided in Excel or Calc can fit a simple curve to a collection of data points. The process begins by right-clicking on the plotted data and, as shown in figure 2.9, a menu pops up that offers the 'Add Trendline' option.

Figure 2.10 shows the pop up menu that appears, where the user needs to select the type of plot that is being estimated. In this case, the data are from a second-order polynomial thus the 'Polynomial' option with 'Order 2' is selected. At the bottom are options which, when turned on, will print the equation and R^2 (error) value of the estimation. The result is shown in figure 2.11 where a plot of the estimate and the equation are shown. This matches the equation that generated the data.

Usually, the lab will produce data that can be plotted but do not follow an equation perfectly. Basically, the data have noise. Figure 2.12 shows a case in which noisy data are estimated by the trend line. Again 'Polynomial' 'Order 2' was selected as the estimating function and, as can be seen, Excel found a second-order polynomial that mostly fits the data. This should be the best possible polynomial that will fit these data.

The most common error in using Trendline is to select the wrong estimating function. Usually, the correct formula type is obtained from the theory. If the theory suggests a log function then the estimated trend line should be a log instead of a polynomial.

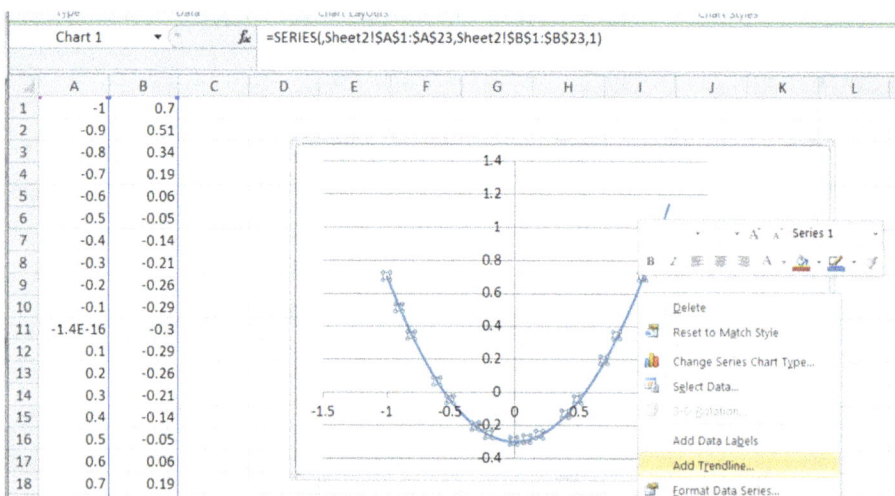

Figure 2.9. The process of creating a trend line starts with a right click on the data.

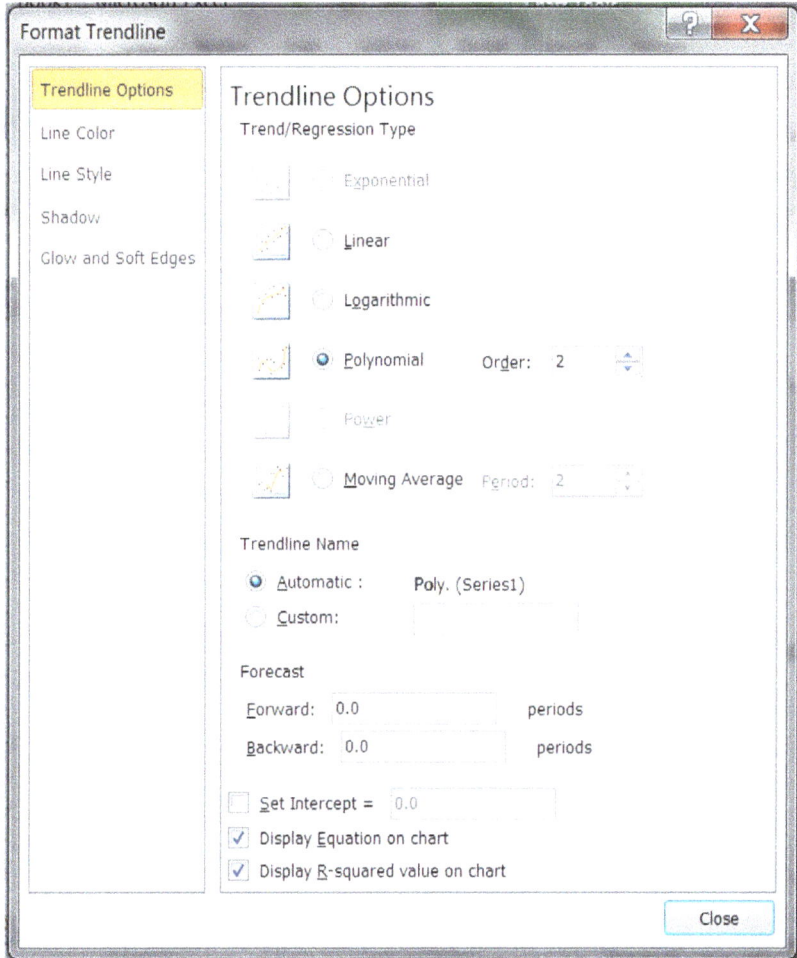

Figure 2.10. The Trendline pop up window in Excel allows the user to select the appropriate functional form.

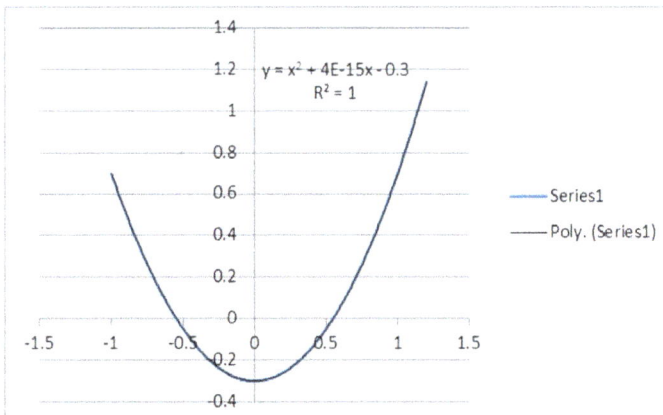

Figure 2.11. The result showing the estimated function and the R^2 value.

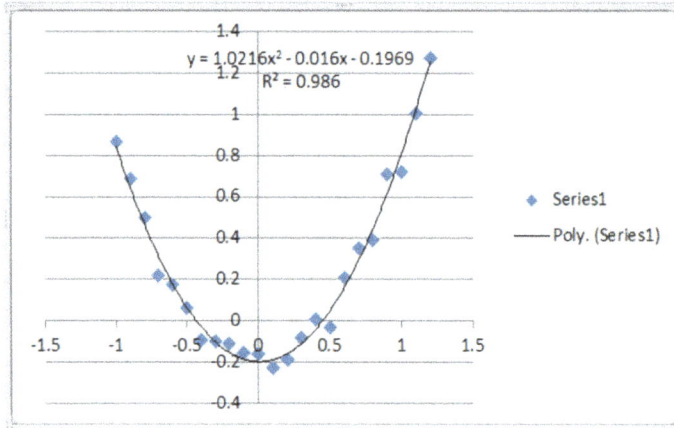

Figure 2.12. The result from noisy data which shows the best polynomial fit and the associated R^2 value.

2.5 Solver

Trendline does work well for the functions it lists, but does not work well for more complicated functions such as a Gaussian (bell curve). For the more complicated functions Excel and LibreOffice Calc offer a Solver function that can estimate the parameters of a function that fits the data.

In chapter 5, for example, the data can be fitted with a Gaussian function with the form,

$$y = Ae^{-(x-\mu)^2/2\sigma^2}, \tag{2.1}$$

where A is the amplitude, μ is the x location of the Gaussian peak and σ is the half width of the peak at half height. For this example $A = 1$ and so the only two parameters are μ and σ. The raw data are shown in figure 2.13 which is created by using $\sigma = 3$ and $\mu = 0.75$; some random noise is added.

In an actual experiment the values of μ and σ are not known and it is the goal of Solver to determine the two values that best fit these data. Using Solver requires a bit more set up work than Trendline. A typical use is shown in figure 2.14 where the raw x and y values are in the first two columns. There are 70 rows of data and this image only shows the first few rows. Column C contains the two variables μ and σ in cells C3 and C5, respectively. Initially, these values are not known and they are set to 1. Column D shows the calculated results using equation (2.1) with the two values of μ and σ from column C. The equation used in cell D2 is shown in line 1 of code 2.1. Column E is the squared error between the measured data (column B) and the calculated data (column D). The Excel command used in cell E2 is shown in line 2 of code 2.1. The difference between the measured and calculated data is squared to remove any negative signs and to accentuate those cases where the difference is large.

Initially, this error is large because the correct values for μ and σ are not known. The final cell G2 is the sum of the errors. The equation for this cell is shown in line 3 of code 2.1. Since all of the squared errors are positive values the only way that cell

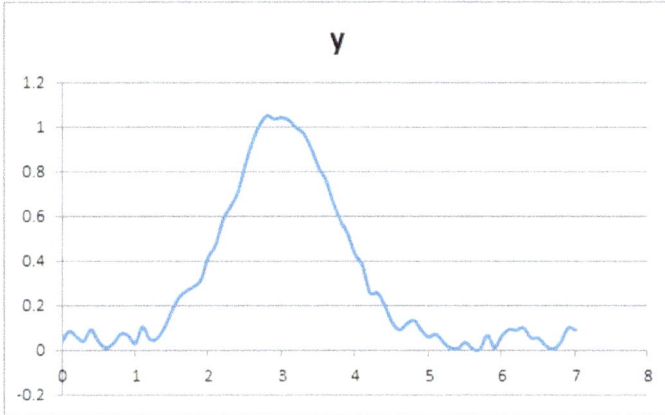

Figure 2.13. Raw data which produce a noisy bell curve.

	A	B	C	D	E	F	G
1	x	y-measured		y-calc	sqerr		
2	0	0.035703	mu	0.606531	0.325845	Error	21.52216
3	0.1	0.084954	1	0.666977	0.338751		
4	0.2	0.060759	sigma	0.726149	0.442744		
5	0.3	0.040543	1	0.782705	0.550804		
6	0.4	0.09263		0.83527	0.551514		
7	0.5	0.038549		0.882497	0.712249		
8	0.6	0.012026		0.923116	0.830085		
9	0.7	0.029446		0.955997	0.858498		

Figure 2.14. The spreadsheet architecture for Solver.

Code 2.1. Commands used in figure 2.14.

```
1   =EXP(-((A2-C$3)^2)/(2*C$5^2))
2   =(B2-D2)^2
3   =SUM(E2:E72)
```

G2 can be zero is if all of the squared errors are zero and this occurs if the calculated and measured data match exactly. Of course, this will not be the case since the data have some noise.

Therefore, the goal is to minimize the value in cell G2 by changing μ and σ. It is possible for the user to manually change these values and keep the changes if the value of G2 is decreased. Basically, Solver will carry out the same action in a much faster manner. Solver is accessed by clicking on 'Data' in the menu and then 'Solver' in the submenu. Figure 2.15 shows the dialog window that appears. In the 'Set Objective' window G2 is entered since this is the cell that is to be minimized.

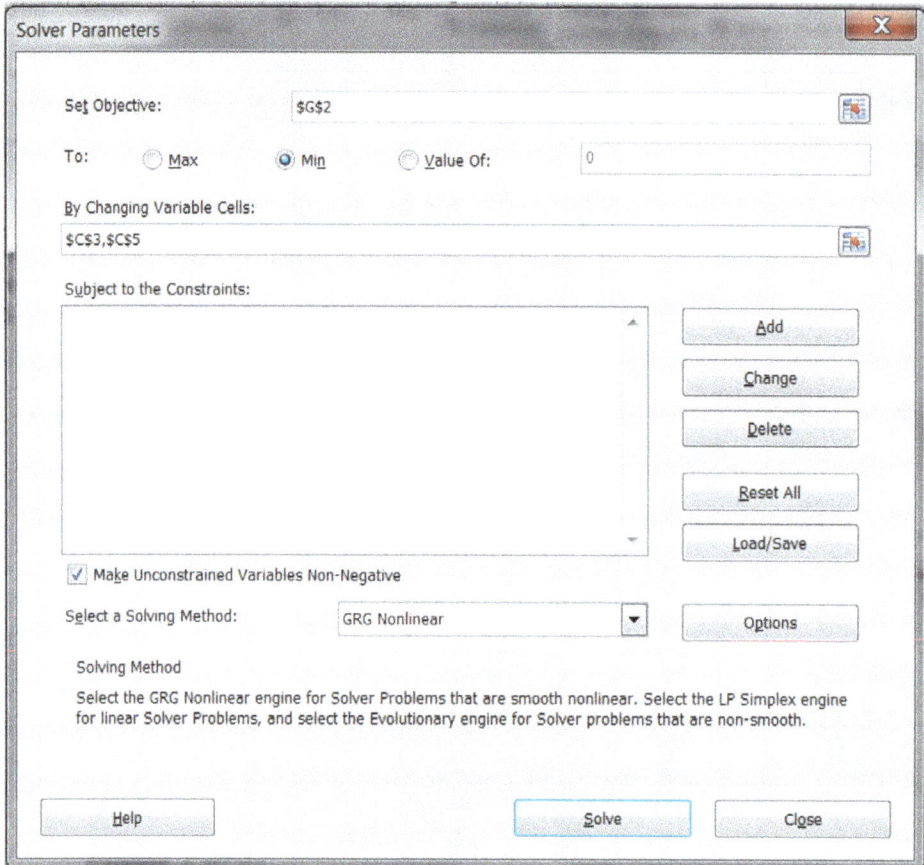

Figure 2.15. The Solver interface.

The desire is that the value of G2 becomes a minimum and so the 'Min' button is selected. Finally, the 'By Changing Variable Cells' window contains the cells that are to be altered which in this case are cells C3 and C5. Finally, the 'Solve' button at the bottom of the window is pressed and Solver computes new values for μ and σ.

In this case the values were computed to be $\mu = 3.00581$ and $\sigma = 0.77699$ which are very close to the values used to generate the data. Had there been no noise then Solver would have recovered the exact values for μ and σ. The final squared error is 0.140. Since the values of μ and σ are now changed the values in columns D and E are also changed. Figure 2.16 shows the new values of column D plotted along with the original data. As seen there is a fairly close match and thus the Gaussian function estimate of the measured data is complete.

Solver is much better suited for problems that Trendline cannot solve. It is important in each case to make sure that the answer provided by the algorithm matches the data. The Solver will return an answer but in some cases the answer may not be sufficiently correct. This is a common issue with these types of algorithms in cases where they cannot home in on a solution or there is something in the data that

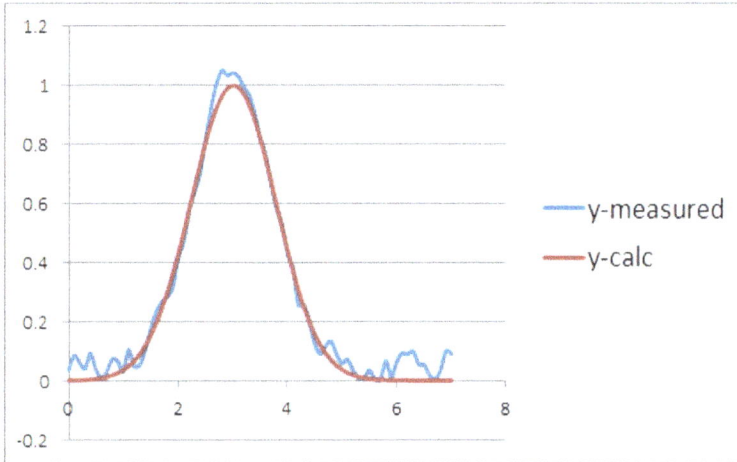

Figure 2.16. Plots of the original data and the Solver estimate.

prevents the algorithm from finding an acceptable solution. If the solution is insufficient then the user needs to identify if there are data points that violate mathematical rules (square root of a negative number, divide by zero, etc) and remove them. If there are a lot of data points then another approach in finding an appropriate solution is to perform the curve fit on a subset of the data.

Chapter 3

Laboratory measurements

3.1 Introduction

In an experiment it is necessary to take measurements. For example, in a velocity experiment it is necessary to measure the distance the object travels and the time it takes to make this journey. However, taking measurements is rarely perfect. If ten people had stopwatches and were asked to measure the elapsed time in an experiment then there would probably be a range of answers even though these people are measuring the same event. The reason is, of course, that each person has a different reaction time.

   This is an example of error in the measurements. It is not possible to say that any one person in the group of ten made the perfect measurement. Maybe one person did but it is not possible to know who it was. Therefore, it is necessary to handle errors in measurements and how they affect our computations. Almost every lab in this course will require recognition of errors and proper computations including these errors. This chapter will explain simple errors and how to handle them in equations.

3.2 Normally distributed data

Most of the data that will be encountered in this course will be considered as normally distributed. This means that a Gaussian function (also known as the *bell curve*) can be used. There are certainly other types of distributions that will not be used in this text.

3.2.1 Measurement

Consider a case in which several measurements of an event are taken. These measurements are denoted as x_i where $i = 1, ..., N$ and N is the number of measurements. As an example, five students are asked to measure the length of a meter stick and their measurements are 1.01, 0.99, 0.99, 1.00 and 1.02. Thus, $N = 5$, $x_1 = 1.01$ m, $x_2 = 0.99$ m, $x_3 = 0.99$ m, $x_4 = 1.00$ m and $x_5 = 1.02$ m.

doi:10.1088/978-1-6270-5628-1ch3

Of course the accepted value of the meter stick is 1.00 m but it is also possible that the stick is a little too short (due to wear and tear) or a little long (because there is extra wood beyond the meter marks on the stick). Even though it is convenient to have a meter stick be exactly one meter long the stick may actually differ from this length. There are two calculations that will aid us in understanding what the measurements mean. These are the *average* (or *mean*) and the *standard deviation*.

3.2.2 Average

The average is simply the addition of the values divided by the number of samples:

$$\bar{x} = \frac{1}{N} \sum_{i=1}^{N} x_i. \tag{3.1}$$

The variable $\bar{x}$ is used to represent the average of all of the x values. It is also common to use the Greek symbol mu, μ, to represent the average.

In the meter stick example the average is $\bar{x} = 1.002$ m. Does this mean that our meter stick is too long? Possibly. There were only five measurements and if 100 people made the same measurements it is still possible that the average would be closer to 1.00. It should also be noted that the average value is off of the accepted value by 0.2%. None of the measurements were to an accuracy of three decimal places and so it can be stated that the average length is 1.00 m within the accuracy of the measured values.

3.2.3 Standard deviation

The average is well known but does not provide enough information. For example the average of the three numbers 3, 4, 5 is 4 and the average of another three numbers 2, 4, 6 is also 4. They have the same average but the second case has a larger variation in the values. The *standard deviation* is the measure of this spread.

The standard deviation is represented by the lowercase Greek letter sigma and is computed by

$$\sigma = \sqrt{\overline{(x_i - \bar{x})^2}} = \sqrt{\frac{1}{N} \sum_{i=1}^{N} (x_i - \bar{x})^2}. \tag{3.2}$$

Usually throughout the course, a spreadsheet will be used to calculate the average and standard deviations.

The standard deviation of the meter stick example is $\sigma = 0.013$. A larger standard deviation means that the data have a larger variation in the measured values. If all measurements were the same then the standard deviation would be exactly 0.

3.2.4 Normal distribution

So, what do the average and standard deviation really tell us? Consider a case in which a large number of measurements is made of a certain object and the average is 1.3 and the standard deviation is 0.6. For a *normal distribution* the data would be somewhat like that shown in figure 3.1. This is a normal distribution which is also called a *Gaussian distribution* or *bell curve*.

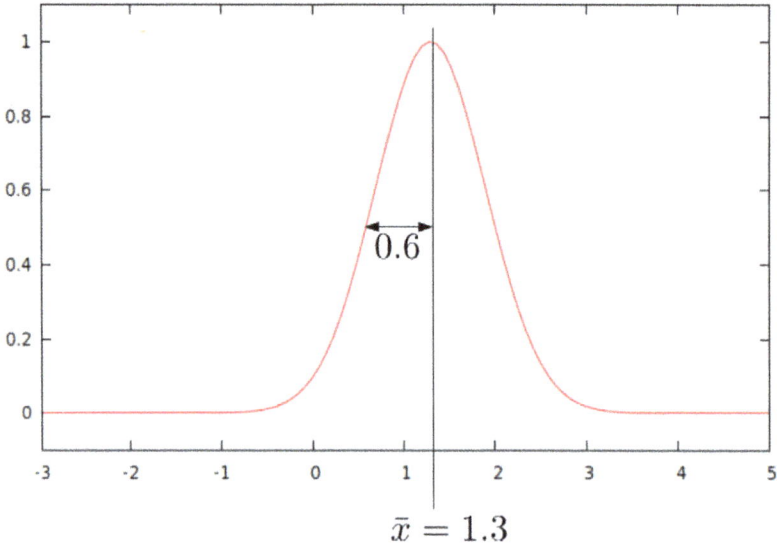

$$\bar{x} = 1.3$$

Figure 3.1. A normal distribution with $\bar{x} = 1.3$ and $\sigma = 0.6$.

This curve is generated solely from the average and standard deviation using

$$y = \exp\left(-\frac{(x - \bar{x})^2}{2\sigma^2}\right),$$ (3.3)

which can also be written as,

$$y = e^{-\frac{(x-\bar{x})^2}{2\sigma^2}}.$$

In this example it is seen that the center of the curve is at $\bar{x} = 1.3$. The x-axis location of the center of the curve is the average of the data. At half height ($y = 0.5$ in this example) the half width of the curve (from the center to the edge) is 0.6 which is the standard deviation.

3.2.5 Working with excel

In almost all the labs a spreadsheet will be used as both a data repository and a calculator. This is important since in some labs the user may have to average over several hundred measurements.

3.2.5.1 Entering data
The data for the meter stick example are typed into spreadsheet cells as shown in figure 3.2.

3.2.5.2 Computing the average within a spreadsheet
Spreadsheets have many functions of which a few will be very useful in this lab. The first function is to compute the average and the function name is AVERAGE. Figure 3.3 shows how this function is entered. The cursor is placed on cell A7 and the

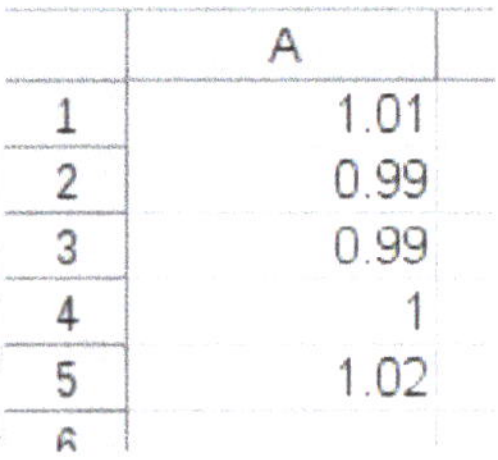

	A
1	1.01
2	0.99
3	0.99
4	1
5	1.02
6	

Figure 3.2. Data in a spreadsheet.

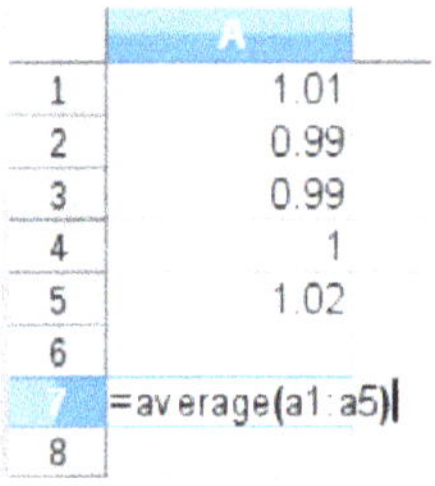

1	1.01
2	0.99
3	0.99
4	1
5	1.02
6	
7	=average(a1:a5)|
8	

Figure 3.3. Computing the average.

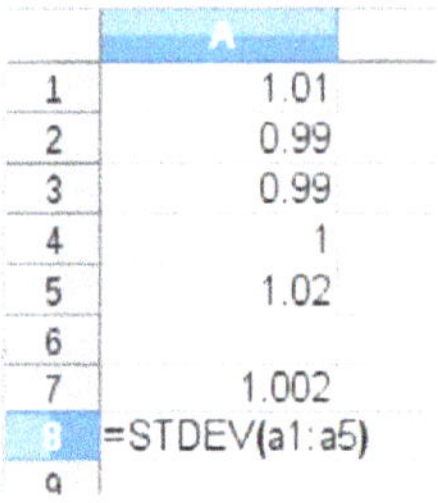

1	1.01
2	0.99
3	0.99
4	1
5	1.02
6	
7	1.002
8	=STDEV(a1:a5)
9	

Figure 3.4. Computing the standard deviation.

user types in =AVERAGE (A1:A5). The equal sign is used to indicate that a function is being created rather than just a word 'average'. The argument to the function is within the parentheses. In this case the average is over all values starting with cell A1 and ending with cell A5. Once the user presses Enter, the function will be replaced with the actual average value. If any of the values in the original five measurements are changed then the average will automatically be updated.

3.2.5.3 Computing the standard deviation within a spreadsheet
Computing the standard deviation is just as easy using the function STDEV as shown in figure 3.4. Once again when the user presses Enter the function will be replaced by the actual value and if the data are updated then the standard deviation will automatically be updated.

3.2.6 Number of σ

In a normal distribution specific percentages of the measurements will fall within specified distances from the average as shown in figure 3.5. 34.1% of the

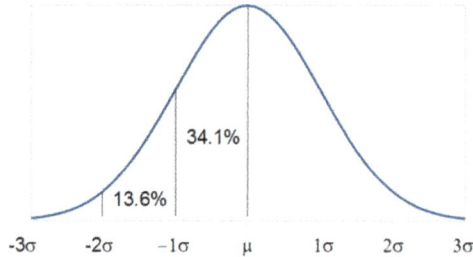

Figure 3.5. Sigma bands in a normal distribution.

measurements are at a distance of less than 1σ to the left of the average. Another 34.1% of the measurements are within 1σ to the right of the average. This means that if 100 measurements were taken then about 68 of these measurements should be within a distance of 1σ from the average. The next band shows the value of 13.6% up to a distance of 2σ from the mean. Again this is on both sides of the average. This means that 95.4% (68.2% + 13.6% + 13.6%) of the measurements will fall within a distance of 2σ from the average. Likewise, 99.6% of the measurements should fall within a distance of 3σ from the average.

Consider an experiment intended to determine a value that is already known, such as calculating gravity (see chapter 16). The experimental value will have a value with an uncertainty represented by $g_e \pm \sigma_{g_e}$ where g_e is the average of the experimental measurements and σ_{g_e} is the standard deviation of those same measurements. If the distance between the average value and the accepted value is less then one standard deviation then the experimental value and the accepted value are considered to be equivalent within the precision of the experiment.

Another type of experiment may produce a value for a variable x through a computation. This would also have an uncertainty associated with it thus providing $x_c \pm \sigma_{x_c}$ where the subscript 'c' indicates that this is a computational value. The experiment may also measure this same variable providing an experiment value $x_e \pm \sigma_{x_e}$ which also has an uncertainty. The two are considered equivalent if $|x_e - x_c|$ is less than the larger of $(\sigma_{x_e}, \sigma_{x_c})$.

3.2.7 Types of error

The two main types of error are *random error* and *systematic error*. Random error is just that. It may arise from the reaction time of the humans or detectors or in the ability of the detector to make a correct measurement. Systematic error usually arises from a specific cause. For example, the accelerometer in a phone may display values for gravity consistently near 9.5 instead of 9.8. This bias is a systematic error.

Random error tends to affect the standard deviation but it may not significantly affect the average. If the random error is larger then the standard deviation is larger. Systematic error, on the other hand, tends to affect the average. If there is a large random error then the Gaussian curve will widen and if there is a systematic error then the Gaussian curve will shift to the left or right.

3.2.8 Accuracy and precision

These types of errors are also related to the accuracy and precision of the measurements. The *accuracy* corresponds to the closeness of the average to the actual value. The *precision* corresponds to the standard deviation.

Consider a simple case in which a person is throwing darts at a dartboard. If all of the shots are very close together then there is a high precision (low standard deviation). It doesn't matter if the shots are close to the center of the target or not. If a partner throws a bunch of darts and they are all around the center of the target but the distance between the darts is large then this is a case of high accuracy. In other words, the average is close to the actual value (the center of the target) even though the standard deviation is high.

Of course, in labs high accuracy and high precision are desired. If the value of σ is small then there is a high precision. If $\bar{x}$ is close to the accepted value then there is high accuracy. The latter is measured by comparing the distance of the accepted value to $\bar{x}$ with respect to σ.

If the accepted value is within 1σ of $\bar{x}$ then the measurements are considered valid. If the accepted value is within 2σ then the measurements are suspect. If the accepted value is greater than 3σ from the average then the validity of the data is drawn into question. It may be that valid data cannot be obtained for this experiment.

3.2.9 Percent difference

This concept is expressed as the *percent difference* between an accepted value and the measured value. Using O to represent the observed value (what was measured in the lab) and A to represent the accepted value then the percent difference is computed by[1]

$$\%D = \frac{|O - A|}{A} \times 100\%. \tag{3.4}$$

Consider a case of measuring gravity. The expected value should be 9.8 m s^{-2} but the average of our measurements may be 9.9. Then $A = 9.8$ and $O = 9.9$. The percent difference is,

$$\%D = \frac{|9.9 - 9.8|}{9.8} \times 100\% = 1.02\%.$$

This means that the measured value deviates from the accepted value by 1.02%. In real world measurements this is a small percent difference.

---

[1] In some publications this is called the *percent error* and the *percent difference* is computed by $\%D = \frac{|O-A|}{1/2(O+A)} \times 100\%$.

3.3 Propagation of errors

In experiments, the *independent variables* are measured and the values for the *dependent variables* are computed. Consider a simple case of velocity where the independent variables are time, t, and distance, x. The dependent variable is v and these are related by $v = \frac{x}{t}$. This process is slightly more complicated if there are errors involved. Here the terms uncertainty, error and standard deviation are the same entity. A known error for a variable is indicated as

$$x \pm \sigma_x.$$

In the example both the time and distance measurements have an error and so there are $t \pm \sigma_t$ and $x \pm \sigma_x$ and the question is how to go about calculating $v \pm \sigma_v$? The *propagation of errors* is the practice of calculating a dependent variable when the independent variables have errors.

3.3.1 Addition and subtraction

Consider a simple case in which x and y are the independent variables and z is the dependent variable. In this case these variables are related by a simple addition, $z = x + y$. However, in this case, there are $x \pm \sigma_x$ and $y \pm \sigma_y$ and the task is to determine $z \pm \sigma_z$.

The calculation is performed in two parts. The first is $z = x + y$ and the second is

$$\sigma_z = \sqrt{\sigma_x^2 + \sigma_y^2}. \tag{3.5}$$

For subtraction the values are $z = x - y$ and $\sigma_z = \sqrt{\sigma_x^2 + \sigma_y^2}$ and as can be seen the calculation of σ_z is still using the plus sign.

Consider an addition case were $x \pm \sigma_x = 5.0 \pm 1.0$ and $y \pm \sigma_y = 6.0 \pm 0.5$. Then $z = x + y = 11.0$ and according to equation (3.5) $\sigma_z = 1.118$. The result is shown in figure 3.6. Here the blue line corresponds to x, the red line corresponds to y and the green line corresponds to z. It is easily seen that the width of the z curve is wider than the other two. This is confirmed by noting that $\sigma_z > \sigma_x$ and $\sigma_z > \sigma_y$. The error of the result is larger than the individual errors of the independent variables.

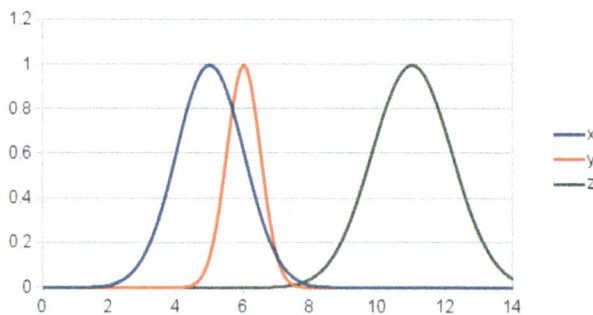

Figure 3.6. The addition of two variables with error.

3.3.2 Multiplication and division

Propagation of errors with multiplications (and divisions) requires the use of *percent sigma*. This is the percentage of error with respect to the value and for a variable z is computed by

$$\%\sigma_z = \frac{\sigma_z}{z} \times 100\%. \tag{3.6}$$

Likewise the conversion from percent sigma to sigma is

$$\sigma_z = \frac{\%\sigma_z}{100} \times z. \tag{3.7}$$

For the case of a simple multiplication ($z = xy$) the percent sigma is calculated by

$$\%\sigma_z = \sqrt{\left(\%\sigma_x\right)^2 + \left(\%\sigma_y\right)^2}. \tag{3.8}$$

For the case where $z = ax$ and a is a constant percent sigma is calculated by

$$\sigma_z = a\sigma_x. \tag{3.9}$$

For the case of a simple division ($z = x/y$) the computation for the value of σ_z is the same as with the multiplication case. Equation (3.8) is the same for multiplication or division excepting the value of z.

3.3.3 Functions

Functions such as trigonometric functions, exponents or logarithms are handled by a more brute force approach. Consider a case were $z = \sin(x)$ and the independent variable has an error, $x \pm \sigma_x$. In this case the sines of $x + \sigma_x$, x and $x - \sigma_x$ are calculated. The error σ_z is the difference from an extreme value to the middle value. The calculation of σ_z is

$$\sigma_z = \frac{|\sin(x + \sigma_x) - \sin(x - \sigma_x)|}{2}. \tag{3.10}$$

3.3.4 Any function

It is possible to derive the form for the error calculation for any function of independent variables. Consider a function $z = f(x, y, \ldots)$ where x, y and the variables that follow are independent. The calculation of the error is,

$$\sigma_z = \sqrt{\left(\frac{\partial f}{\partial x}\right)^2 \sigma_x^2 + \left(\frac{\partial f}{\partial y}\right)^2 \sigma_y^2 + \ldots}. \tag{3.11}$$

This equation requires that all the variables (x, y, ...) be independent of each other.

Chapter 4

Train preparation

One of the restrictions on these labs is that they must be suitable for students working at home. Thus, all materials must be available at local stores or hobby shops. Several experiments will need to use motion with very little friction. In school labs this is often accomplished with the use of an air track but these are not easily available and are expensive. So the solution here is to use model trains which can provide low friction motion with minimal cost. However, this track will be used in several experiments so the expense can be averaged over multiple labs. This chapter explains the construction of the track system which will be used in several labs.

While this system does not provide frictionless motion it does provide motion with friction that is low enough to be acceptable. Inaccuracies in collecting data will overshadow the inaccuracies caused by this small amount of friction.

4.1 Materials

There are several scales for model trains with the two most popular being N and HO. The HO scale train is recommended even though this means a slight increase in cost. The reason is that the HO scale trains demonstrate significantly lower friction than the N scale trains. For these experiments only two train cars are required. There is no need to purchase an engine car or the power supply.

The materials needed to construct the train system are:
- at least three feet of HO scale train track,
- nails that are designed for train track,
- two gondola (or flatbed) rail cars,
- a thin sheet of plywood roughly measuring 4 feet by 1 foot,
- strips of Velcro, and
- one medium sized binder clip.

4.2 Construction

The construction of the track is straightforward as shown in figure 4.1. The track is nailed to a board so that it remains level. In this example a flexible track was used

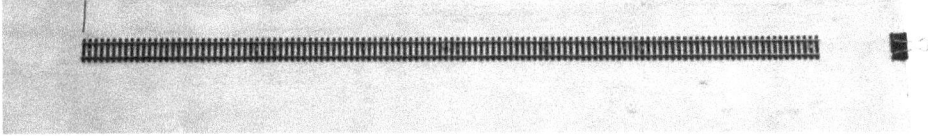

Figure 4.1. Train track is nailed to a piece of wood to maintain a straight and level run. A binder clip is attached to the wood to provide a smooth edge for strings that will hang over the edge of the wood.

Figure 4.2. A gondola car with a smart phone.

because it provided a three foot long piece without any junctions so the rails are continuous for the entire run. The downside is that flexible track can also curve. A meter stick was placed next to the track to ensure that the track was straight before it was nailed to the board.

In several experiments a force will be applied to the train car through a string or fishing line. The other end of the string will be attached to a weight that will hang down from the board. The string, however, bends over the edge of the wood which will add friction. Thus, a binder clip is attached to the end of the wood to provide a smooth surface for the string to run against.

4.3 Train cars

Most experiments will use one train car with the exception of the experiments involving collisions. The train car will need to hold the smart phone and some small weights. Thus a gondola (shown in figure 4.2) is the desired car type although a flatbed may also work.

The gondola is preferred for the collision experiments as well. The trains have plastic couplers which tend not to survive these collisions. A better solution is to attach strips of Velcro to the ends of the cars so that they stick together upon collision. The gondolas offer a vertical flat surface to attach these strips whereas the flatbeds do not.

4.4 The finished system

The finished system will have a piece of track that is secured to a piece of wood. The wood can also be marked to denote specific locations along the track. This train system is the most expensive component of the labs but it will be used in labs 4, 5 and 7.

Part II

13 Labs

Chapter 5

Lab 1: acceleration of an elevator

5.1 Educational goal

The goal of this lab is to relate time-dependent acceleration to displacement.

5.2 Materials

Materials needed for this lab are:
- acceleration logger app,
- an elevator, and
- a spreadsheet.

5.3 Theory

Given an acceleration function dependent on time, $a(t)$, the computation of the time-dependent velocity function is

$$v(t) = \int_{t_1}^{t_2} a(t)\, \mathrm{d}t, \tag{5.1}$$

where t_1 and t_2 are the beginning and ending times of the motion defined by $a(t)$. Likewise, the time-dependent position function can be determined by

$$x(t) = \int_{t_1}^{t_2} v(t)\, \mathrm{d}t. \tag{5.2}$$

In this lab the acceleration data will be collected during an elevator ride and from this the height of the elevator ride will be determined. There are three different components to the acceleration function for this experiment. There is the acceleration as the elevator gets up to speed, $a_1(t)$, from times t_1 to time t_2. There is a constant velocity segment, $a_2(t) = 0$, where the elevator travels between floors from times t_2 to t_3 and there is the deceleration, $a_3(t)$, from t_3 to t_4 as the elevator slows to a rest. Therefore, the velocity function is actually computed by

$$v(t) = \int_{t_1}^{t_2} a_1(t) \, dt + \int_{t_2}^{t_3} a_2(t) \, dt + \int_{t_3}^{t_4} a_3(t) \, dt. \qquad (5.3)$$

The position function is computed likewise in a segmented approach.

The ideal acceleration function would be that shown in figure 5.1 which shows the acceleration as a function of time. This graph shows data before and after the ride as well and these segments will need to be removed before analysis is performed. The three segments are the three distinct portions of the ride: acceleration, constant velocity and deceleration.

In actuality, the collected data will be more like those shown in figure 5.2, which includes noise and vibrations inherent in the elevator. Still, the functional form of the acceleration is present and can be estimated. This lab will collect data, estimate the functional form of the accelerations and then estimate the height that the elevator traveled.

5.4 Procedure

The only equipment needed to collect data is a smart phone with an app that logs acceleration data and an elevator. The steps are:

1. Place the smart phone on the floor of the elevator. Do not hold it in your hand while collecting data since it is not possible to hold the arm perfectly steady.
2. Turn on the accelerometer.

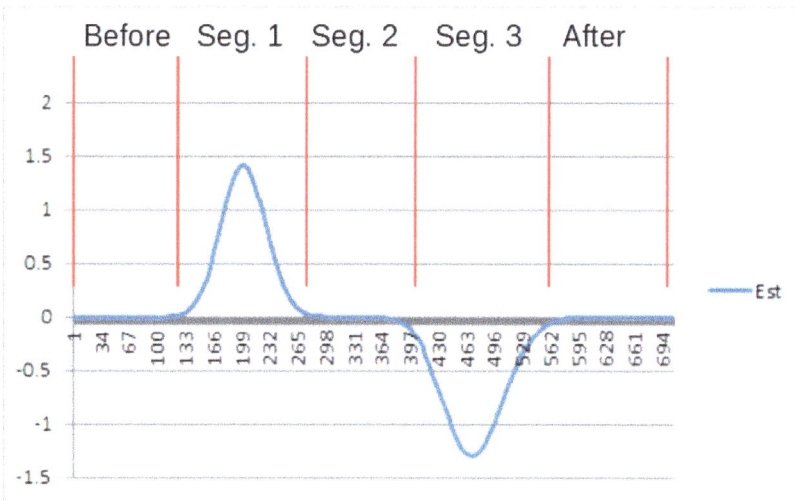

Figure 5.1. The ideal acceleration for an elevator. There are three segments during the ride and a segment before and after.

5-4

Figure 5.2. The actual acceleration data for an elevator from the third channel of the accelerometer.

3. Ride the elevator up two or three stories.
4. Turn off the accelerometer and send the data to a computer with a spreadsheet for analysis.

It may be necessary to perform a few trials to obtain one with sufficiently smooth data.

5.5 Analysis

In the analysis, one will first need to isolate the data for segments 1 and 3 and then estimate the functional form of these segments. Segment 2 will be considered to have constant velocity and so no estimation is required. Once the functional forms are in-hand the last step is to compute the height of the elevator ride.

To estimate the acceleration with a function it is first necessary to isolate the acceleration portion of the data as shown in figure 5.3 [1]. Note that the values along the x-axis still correspond to time.

Figure 5.1 shows segment 1 as a Gaussian function. This will be cumbersome to handle and therefore the data will be fitted with a second-order polynomial. With the amount of noise in the data it is believed that a polynomial fit will not introduce significant errors compared to a Gaussian fit. The curve fit is performed using the Trendline function in the spreadsheet (see section 2.4). An example with the estimation is shown in figure 5.4 which fits segment 1 with $a(t) = -0.67t^2 + 4.05t - 5$.

5.5.1 Functional form

The acceleration is estimated by a quadratic function so it is of the form

$$a(t) = \alpha t^2 + \beta t + \gamma, \qquad (5.4)$$

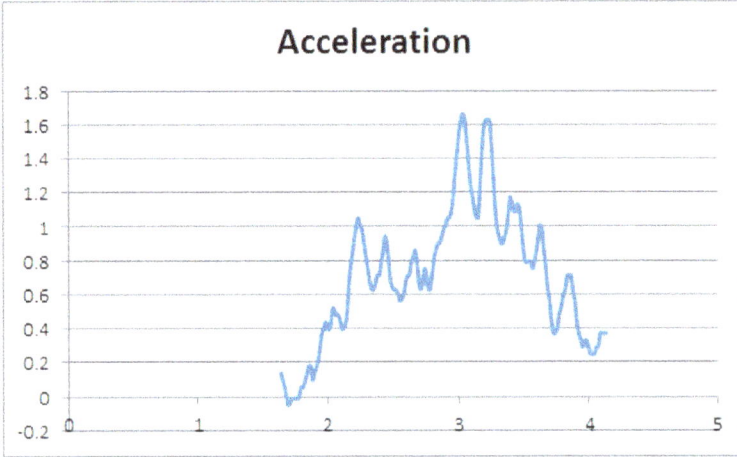

Figure 5.3. The data from the first acceleration.

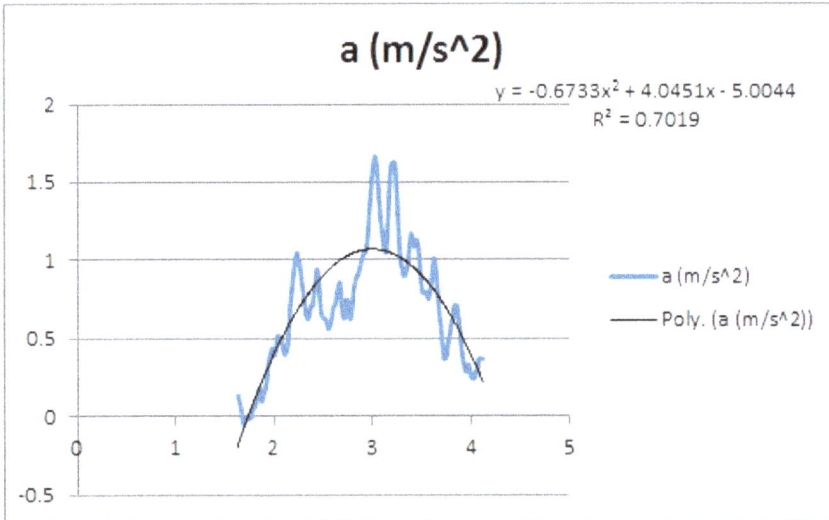

Figure 5.4. The data from the first acceleration and an estimation of these data with a quadratic function.

where α, β and γ are constant and to be determined. The velocity function becomes

$$v(t) = \int a(t)\, dt = \frac{1}{3}\alpha t^3 + \frac{1}{2}\beta t^2 + \gamma t + c_1 \qquad (5.5)$$

and the position function becomes

$$y(t) = \int v(t)\, dt = \frac{1}{12}\alpha t^4 + \frac{1}{6}\beta t^3 + \frac{1}{2}\gamma t^2 + c_1 t + c_2. \qquad (5.6)$$

The constant c_1 is determined through boundary conditions. At the beginning of segment 1, $v = 0$, and therefore from equation (5.5),

$$c_1 = -\frac{1}{3}\alpha t_0^3 - \frac{1}{3}\beta t_0^2 - \gamma t_0. \qquad (5.7)$$

In this example the acceleration starts at $t_0 = 1.76$ s and so $c_1 = 3.77$ m s^{-1}. Likewise, the constant c_2 can be computed by noting that at the beginning of segment 1 $y = 0$ which produces $c_2 = -2.01$ m. The process is repeated for segment three and for the example data $\alpha = 0.906$ m s^{-4}, $\beta = -13.55$ m s^{-3}, $\gamma = 49.5$ m s^{-2}, $c_1 = -116.7$ m s^{-1} and $c_2 = 204.6$ m.

The height of the elevator ride is determined in three parts. First the height of segment 1 is determined by equation (5.6). The velocity at the end of this segment is determined by equation (5.5) since it is the initial velocity of the second segment. Since this segment has a constant velocity its height is $y_2 = v_2 t_{23}$, where v_2 is the velocity at the end of the first segment and $t_{23} = t_3 - t_2$ is the duration of the second segment. The height of the third segment is computed in a fashion similar to segment 1 except that there is an initial velocity for this segment. The total height is the sum of the three heights and for this example the height was estimated to be 8.1 m which is well within the average of 4 m per story.

5.6 Procedure

The procedure is simple. The acceleration logger is turned on and the phone is placed on the floor of the elevator. The elevator is then ridden upwards for two or three stories. The acceleration logger is then turned off and the data are sent to a spreadsheet for analysis. A single story elevator ride may be too short causing segments 1 and 3 to overlap. Rides of more than three stories just lengthen segment 2 which adds no value to the experiment.

5.7 Analysis

The analysis is also straightforward. The data are viewed in a spreadsheet and the correct channel of data is isolated and it should appear similar to figure 5.2. If the phone is upside down then the graph will also be upside down, but this can be easily corrected in the spreadsheet.

The boundaries of the five segments are identified and data from segments 1 and 3 are extracted to their own spreadsheet pages. For each of these the second-order polynomial Trendline fit is applied providing values for α, β and γ. Constants c_1 and c_2 are then computed as outlined in section 5.5.1.

Calculations for the heights of the segments are then performed and the total height is the sum of the three. It should be noted that commercial buildings have an average story height of slightly less than 4 m and the results should be within this range. If the actual heights of the stories are known then it is possible to compute the percent difference between the actual and measured values.

5.8 Possible problems

Possible problems with this lab are:
- The phone should be placed on the floor or a solid surface. If the phone is held in the hand then there will be additional noise from the movements of the arm.
- If the elevator ride is jerky then there will be a lot of noise. The only solution is to find a different elevator.

Bibliography

[1] Kinser J M 2015 Relating time-dependent acceleration and height using an elevator *Phys. Teacher* accepted

Chapter 6

Lab 2: independence of x and y

6.1 Educational goals

The purpose of this lab is to demonstrate that motions in the x and y planes are independent.

6.2 Materials

Materials needed for this lab are:
- a video camera and video analysis software (see section 1.3.2) or a burst camera,
- a basketball (or similarly large object),
- a safe site at least two stories high, and
- a meter stick or measuring tape.

6.3 Theory

The equation of motion for an object dropped from a height h_0 is,

$$h = h_0 + v_{0y}t + \frac{1}{2}gt^2, \tag{6.1}$$

where v_{0y} is the initial velocity in meters per second, h is the height in meters, $g = -9.8\,\mathrm{m\,s^{-2}}$ and t is the time in seconds.

The vertical motion has acceleration due to gravity, if there is horizontal motion it is assumed to be at a constant velocity. Therefore, the horizontal motion is described by,

$$x = x_0 + v_{0x}t, \tag{6.2}$$

where x_0 is the initial horizontal position and v_{0x} is the initial horizontal velocity.

doi:10.1088/978-1-6270-5628-1ch6

Since the motion in the vertical and horizontal dimensions are independent the horizontal velocity does not affect equation (6.1). Therefore objects with no initial vertical velocity will fall at the same rate independent of the initial horizontal velocity. This, of course, applies to objects that are not impeded by air resistance.

6.4 Procedure

This lab is performed with two experiments. The *control* experiment is to drop an object from a large height and measure the time to reach the ground. The second experiment will add a horizontal velocity to the first experiment. If the motions in the vertical and horizontal dimensions are independent then the objects should reach the ground in the same amount of time.

Unfortunately, h_0 needs to be large which adds an element of danger to this experiment. If an object were allowed to drop for three seconds then according to equation (6.1) the object would fall 44 m which is about an 11 story building. Furthermore, by the time the object reaches the ground it could have a velocity that exceeds $30 \, \mathrm{m \, s^{-1}}$ which is about 67 mph. An object that is dropped two stories will reach the ground in 1.25 s. Reality dictates that these experiments must be performed in a very short time.

Figure 6.1 shows a system where a basketball can either be dropped from a height h_0 or pushed off of a horizontal surface at the same height. The basketball is a good object to use because it will be large enough to clearly see in the video images and it will not be subject to significant air resistance.

At least three students should be used for this experiment. One student is responsible for dropping the basketball, one student is responsible for collecting the data and the third student is used for safety. It is the responsibility of the third student to make sure that there are no other people in the area where the basketball is landing. Even from a height of two stories the ball could approach 25 mph by the time it reaches the ground.

The data are collected using either a video or burst camera app in the smart phone. If a video is captured then the students will also need simple video analysis software that will allow the user to step through the video one frame at a time with a

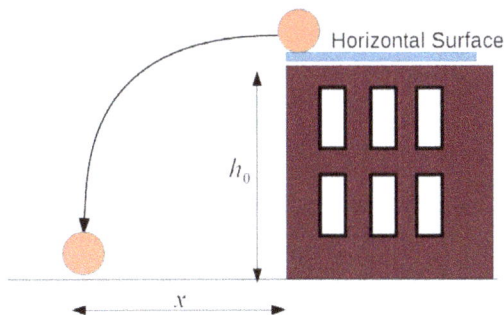

Figure 6.1. Schematic for the experiment. A basketball can be dropped from a height h_0 or pushed off a horizontal surface at the same height.

time stamp. Burst camera apps generally encode the time stamp in the file name and thus are a better choice.

In the first experiment the basketball is dropped from a height h_0 without any initial vertical velocity $v_{0y} = 0$ or any significant horizontal velocity $v_{0x} \approx 0$. The motion from the start to the end of the drop is captured by the camera (video or burst). In the second experiment the ball is pushed along a horizontal surface so that the ball also has a significant horizontal velocity. Again this motion is captured by the camera. Both experiments may be repeated multiple times to ensure confidence in the measurements.

The measured data for each trial of the first experiment are the start time (when it starts to fall) and the end time (when it lands). In a single trial the time at which the ball begins to drop, t_0, and when the ball lands, t_1, are extracted from the camera images. It is expected that the exact moment when the ball begins and when it ends its descent will be between camera frames and thus these exact moments will have to be estimated with an error (time between frames). If multiple trials are used then the average drop time is used along with its standard deviation. The drop time is therefore represented by $t \pm \sigma_t$. The horizontal distance $x \pm \sigma_x$ should also be measured.

For the control experiment the measured variable is $t_c \pm \sigma_{t_c}$. For the second experiment the measured variables are $t_e \pm \sigma_{t_e}$ and $x \pm \sigma_x$.

6.5 Analysis

The measured time for the first experiment is represented as $t_c \pm \sigma_{t_c}$ and the time for the second experiment as $t_e \pm \sigma_{t_e}$. If the vertical and horizontal motions are independent then these two times should be equivalent within their given errors. Thus, the hypothesis of independent motion is supported if the values of t_c and t_e fall within 1σ of each other.

Two other computations can be made as well. The first is to compute the height of the building $h_0 \pm \sigma_{h_0}$ using equation (6.1). These results can be confirmed if the height of the building is known, or they can at least be determined to be reasonable. For example, the average height of an office story is 3.9 m and if the object is dropped two stories then the computed value of h_0 is reasonable if it is somewhat close to 7.8 m.

The other computation is to determine the initial horizontal velocity for the second experiment using equation (6.2). This value can also be confirmed by comparing it to the velocity calculated by $v_x = x/t$.

6.6 Possible problems

There are a few aspects of this lab that can be problematic.
1. The object must be large enough to be visible on the video. For example, a golf ball will be difficult to see especially if the background has similar colors. A basketball is much easier to see in the video frames.
2. The estimates of σ_t must not be too small. There are two sources for σ_t. The first arises because it may not be possible to determine from the images

the exact start and stop time. The second comes from computing the standard deviation from multiple trials. Students should take all causes into account when determining σ_t. If the value of σ_t is too small then the computation of the number of sigmas will be artificially large.

3. The launching surface must be horizontal. If it is slanted then there will also be an initial vertical velocity which will be different in each trial.

Chapter 7

Lab 3: tension

7.1 Educational goals

The goal of this lab is to explore forces in equilibrium within a two-dimensional architecture.

7.2 Materials

The materials needed for this lab are:
- a vertical surface (such as a bulletin board),
- springs (at least two),
- various weights,
- string and items that will attach the strings to the vertical surface (such as thumbtacks),
- a meter stick, and
- a digital camera.

7.3 Theory

The force, F, required to stretch a spring an additional distance, Δx, is linearly proportional to that distance:

$$F = -k\Delta x, \tag{7.1}$$

where k is the spring constant.

Figure 7.1(a) shows a simple system in which the a weight, w, hangs from a string that is stretched a distance Δx. Since the system is in equilibrium the force of the spring must equal the weight and therefore

$$k\Delta x = w = mg. \tag{7.2}$$

Figure 7.1(b) shows a two-dimensional system in equilibrium in which the weight is supported by two springs. Since the motions in the x and y directions are

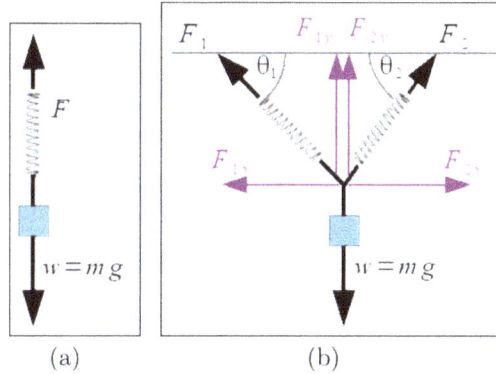

(a)   (b)

Figure 7.1. Schemes for spring forces in equilibrium for one-dimensional and two-dimensional cases.

independent the vertical forces should sum to zero and the horizontal forces should sum to zero. Thus,

$$F_{1x} + F_{1y} = mg \tag{7.3}$$

and

$$F_{1x} = F_{2x}, \tag{7.4}$$

where $F_{1x} = F_1 \cos \theta_1$, $F_{1y} = F_1 \sin \theta_1$ and similar equations exist for F_2.

7.4 Procedure

The ultimate goal is to construct the systems shown in figure 7.1(b) and show that the forces are in equilibrium. Before this is carried out it is necessary to calibrate the springs. So, the experiment has two steps: 1) calibrate the springs and 2) show equilibrium in the two-dimensional system.

7.4.1 Calibration

This step is to the determine the spring constant, k, for each spring that will be used. The first step is to choose a proper spring. If the spring is too strong then the weights will not be able to stretch it to any significant length and thus $\Delta x \approx 0$. If the spring is too weak then k will not be a constant. A weak spring can be identified if the coils have different distances between them when the spring hangs vertically without any weight attached. So, determining the spring constant will require measuring Δx for several different weights within the range of the weights to be used in the experiment.

For each spring several trials are required in which a weight is attached to the bottom of the spring and Δx is measured (figure 7.2(a)). Using equation (7.1) the spring constant k is computed for each trial. A spring of sufficient strength should have a constant value of k for a wide range of weights. The multiple trials for each spring will probably not produce exactly the same value for the spring constant and thus each spring is defined by $k \pm \sigma_k$.

It may not be possible to find springs with a true spring constant. In these cases the calibration is performed after the experiment. The experiment will provide the Δx values for the spring. In the calibration stage the weights are added to the spring until the same Δx value is obtained. These weights are equivalent to the force used in the experiment.

7.4.2 Equilibrium experiment

The system shown in figure 7.2(b) is constructed. String is attached to each end of the springs and the top end of the string for the upper springs is attached to a bulletin board with thumbtacks. A weight is attached to the lower string as shown. The masses of these weights are recorded. A meter stick is placed in the field of view of the camera for distance calibration.

A photograph is taken of this system and it is important that the smart phone be perpendicular to the bulletin board. If the camera is at an angle to this surface then the image will be distorted. In this case the number of pixels per centimeter would not be uniform throughout the image.

Using software such as GIMP the following components are measured:
- the angles θ_1 and θ_2,
- the number of pixels for a specified length of the meter stick, and
- the length of each spring in terms of the number of pixels in the image.

7.5 Analysis

The first computation is to determine the change in the lengths of the springs Δx_1 and Δx_2. The original lengths are measured by simply placing the springs next to the meter stick on the table. The extended lengths are obtained through the photograph. The ratio ρ is defined as the length of the meter stick (in centimeters) to the number of vertical pixels it consumes in the image. Thus the length of a spring is determined

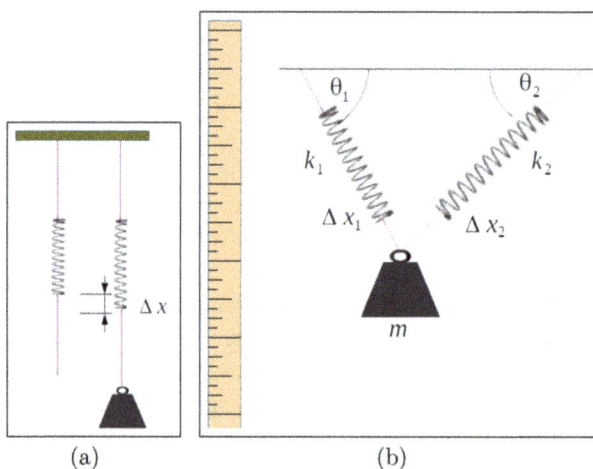

Figure 7.2. Schemes for calibrating the springs and for the two-dimensional experiment.

by multiplying this ratio with the number of pixels that the length of the spring consumes. By subtracting the original lengths the values of $\Delta x_1 \pm \sigma_{\Delta x1}$ and $\Delta x_2 \pm \sigma_{\Delta x2}$ can be computed. Since the spring constants are known from the calibration step the calculations of the forces $F_1 \pm \sigma_{F1}$ and $F_1 \pm \sigma_{F2}$ can be computed use equation (7.1).

The same tool in GIMP that measures lengths in pixels also measures angles to the horizontal. This tool is used to measure the angles θ_1 and θ_2. From the simple geometry described at the end of the theory section, it is possible to compute the horizontal and vertical components of the forces $F_{1x} \pm \sigma_{F1x}$, $F_{1y} \pm \sigma_{F1y}$, $F_{2x} \pm \sigma_{F2x}$ and $F_{2y} \pm \sigma_{F2y}$.

Once the forces are computed it is possible to confirm the equilibrium equations (7.3) and (7.4). Lab weights are usually accurate within 5% of the value imprinted on them and so it is necessary to consider the hanging weight to also have an uncertainty value allowing the weight to be expressed as $w \pm \sigma_w$. Each side of equations (7.3) and (7.4) is computed separately and the equilibrium equations are confirmed if the left and right sides of the equations are within a value of 1σ of each other.

7.6 Possible problems

There are a few aspects of this lab that can be problematic.

1. Determining the length of the spring is difficult because the beginning and ending locations of the coils are not exactly defined. One solution is to mark locations on each end of the spring that act as the beginning and end for the length measurement.
2. Not all springs are suitable for this experiment. If the spring is too strong then it won't stretch. If the spring is too weak then k is not constant. If the coils are not equally spaced when the spring hangs vertically then it is too weak.
3. If the camera is not parallel to the plane of the experiment this will cause distortions in the image and reduce the accuracy of the measurements.
4. The angles θ_1 and θ_2 should not be equal to each other. If $\theta_1 \approx \theta_2$ then $F_{1y} \approx F_{2y}$ and the calculations can be simplified. This is improper to support equation (7.4).

Chapter 8

Lab 4: the Atwood machine

8.1 Educational goals

The goal of this lab is to explore forces in a non-equilibrium system using Atwood's machine.

8.2 Materials

The materials needed for this lab are:
- train track assembly (see chapter 4) and one gondola car,
- a meter stick,
- string or fishing line,
- a small link chain (for an optional experiment),
- scales,
- a smart phone app to log the accelerometer data,
- a smart phone app for orientation, and
- a spreadsheet.

8.3 Theory

This lab has four similar experiments that explore Newton's second law with an Atwood machine. The second law states that

$$\sum_{i=1}^{N} F_i = m_T a, \tag{8.1}$$

where N is the number of forces acting on the mass m_T, F_i are the forces and a is the acceleration of the mass.

These experiments will explore this law with the simple Atwood machine shown in figure 8.1. In the first experiment the only force is provided by gravity acting on m_B. Thus, Newton's law becomes

$$m_B g = m_T a, \tag{8.2}$$

doi:10.1088/978-1-6270-5628-1ch8

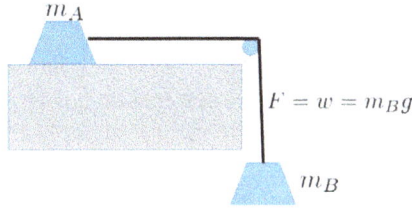

Figure 8.1. The Atwood machine showing the mass m_A being pulled over a frictionless surface by mass m_B.

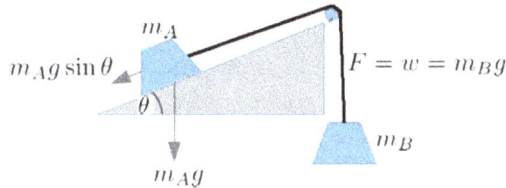

Figure 8.2. The Atwood machine showing the mass m_A being pulled over a tilted frictionless surface by mass m_B.

where m_T is the total mass $m_A + m_B$ and m_A includes the gondola car, the smart phone and any additional weights placed in the car.

The second experiment adds more mass to m_A by putting weights in the gondola car. The theory, however, is the same as in the first experiment. The third experiment adds a second force by tilting the surface by an angle θ as shown in figure 8.2. The acceleration is now resisted by the force acting on m_A due to gravity. Thus, Newton's law becomes

$$m_B g - m_A g \sin \theta = m_T a. \tag{8.3}$$

The fourth experiment replaces m_B with a chain. The chain also has a mass but as it begins to coil on the floor the mass that is participating in the acceleration is decreasing. Thus, the mass m_B is dependent on the distance that the system has traveled. If the system starts with the first link of the chain just touching the floor m_B is linearly dependent on x, the distance that the system travels, which is linked to the acceleration by $x = \frac{1}{2}at^2$. Thus the equation becomes,

$$\frac{m_B t^2}{2X} a = m_T a, \tag{8.4}$$

where m_B is now the mass of the chain, X is the length of the chain and t is the time of the motion.

8.4 Procedure

Four similar experiments are presented. The first will be detailed and the others will follow a very similar procedure with only the modifications being detailed in the text.

8.4.1 Experiment 1

In the first experiment the train is attached to a weight via string that does not stretch (figure 8.3). Some fishing lines may work better than string for this reason. When the weight is released the car is pulled to the right and the acceleration is measured by the smart phone riding on the car.

The mass m_A is determined by summing the measured masses of the gondola car and the smart phone that rides on it. Mass m_B is also measured. The train and mass m_B are connected by a string or fishing line. The car is positioned on the track making sure that the all of the wheels are on the track. The weight is held in place not allowing it to yet pull on the string. The string is positioned so that it will ride over the binder clip when the weight is allowed to fall.

The smart phone is placed on the gondola so that one of the axes of the phone is in line with the direction of motion. The app that logs the acceleration is turned on and the weight is released. This will cause the train to move and it is prudent to catch the train before it falls off the board. The accelerometer app is turned off and the data are sent to a computer for analysis as demonstrated in section 8.5.1.

8.4.2 Experiments 2, 3 and 4

Experiment 2 is a repeat of experiment 1 except that more weight is added to m_A. This can be accomplished by simply placing the extra weight inside the gondola. It is possible that the weight will shift when the car first starts to accelerate and so the weight should be placed where this action is minimized (up against the gondola-wall that is not attached to the string). The additional weight is measured and included in m_A.

Experiment 3 is a repeat of experiment 1 except that the board is tilted at an angle θ as shown in figure 8.2. The angle is measured by using the orientation app on the smart phone. It should be noted that most of these apps do not provide fractions of a degree so the angle is $\theta \pm \sigma_\theta$ where $\sigma_\theta = 1°$. The analysis of this experiment will require equation (8.3).

Experiment 4 is a repeat of experiment 1 except that the weight m_B is replaced by a chain that hangs vertically from the string. The first link of the chain touches the floor before the experiment is started. The length of the chain should not exceed the length that the train can travel thus allowing all of the chain to curl on the floor during the experiment. The analysis of this experiment will require equation (8.4).

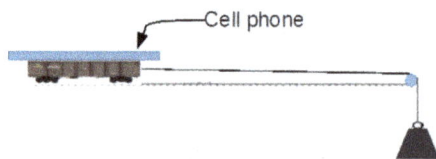

Figure 8.3. The first experiment which shows a smart phone riding on a gondola car which is accelerated by allowing a hanging weight to fall.

8.5 Analysis

As in the previous section only the first experiment will be detailed as the rest are very similar with slight modifications.

8.5.1 Experiment 1

The app that logs the acceleration will create a text file which can be sent to a computer and read by a spreadsheet. An example is shown in figure 8.4 and it should be noted that different apps will present the data in different formats. In this case the first column is a date stamp, the second column is the time measured in nanoseconds and the next three columns are the accelerations of the phone along the three different axes.

In this experiment only one axis of the the phone is needed and that is the one that is parallel along the direction of the motion. This can be easily determined by turning on the app and moving the phone along the axis of motion and identifying which of the three columns of data reacts to that motion.

The app is turned on well before the experiment starts and off long after the experiment ends and so a lot of data are collected that are not part of the experiment. Thus, it is necessary to determine the location of the applicable data. Figure 8.5 shows

	A	B	C	D	E
1	2015/01/15 13:18:44	885697866000	0.068102	-0.027241	10.106298
2	2015/01/15 13:18:44	885727285000	-0.068102	-0.027241	10.18802
3	2015/01/15 13:18:44	885741964000	-0.190685	-0.027241	10.256122
4	2015/01/15 13:18:44	885770437000	-0.027241	-0.027241	10.18802
5	2015/01/15 13:18:44	885784841000	0.068102	0	10.106298
6	2015/01/15 13:18:44	885799063000	0.027241	-0.027241	10.147159
7	2015/01/15 13:18:44	885813528000	-0.068102	-0.027241	10.18802
8	2015/01/15 13:18:44	885827902000	-0.149824	-0.027241	10.215261

Figure 8.4. The first few rows of a typical data collection.

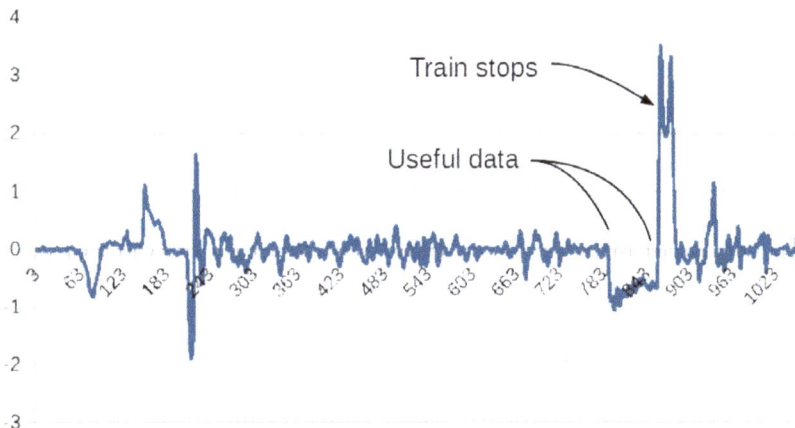

Figure 8.5. The full length of the collected data along the axis of motion.

the raw data along the axis of motion. The horizontal axis is time and the vertical axis is acceleration.

The first feature to notice about these data is that they are upside-down. This occurred because the smart phone was facing in the wrong direction. So, data points above $y = 0$ are negative accelerations. The second feature to notice is the large spike at the right of the chart. This is the sharp deceleration that occurred when the train was stopped. Thus, the data that precede it are the actual experiment. In this experiment the acceleration should be constant during the run and thus there should be a flat section of the trial (which in this case should be below $y = 0$). Since the experiment is not exactly frictionless there will be a slight slant towards $y = 0$ during the trial.

Thus, the actual data start around $x = 800$ and end before the spike. A section of the chart starting at $x = 750$ is shown in figure 8.6. This segment still has too many data but it will show the details of the region that contains the usable data. Near $x = 790$ is the acceleration of the train and near $x = 850$ the sudden stop is seen. Small oscillations are visible because the string acts like a spring. As seen these dampen and for this experiment the oscillations are not invasive.

The usable segment should begin after the motion of the train is settled ($x \approx 795$) and end well before the train is stopped ($x \approx 845$). Once the usable segment is isolated the average and standard deviation of the data are computed. In this case the average is –0.662 and the standard deviation is 0.083. Noting that the phone was facing the wrong direction, the acceleration of the system was measured to be

$$a \pm \sigma_a = 0.662 \pm 0.087 \text{ m s}^{-2}. \tag{8.5}$$

The acceleration is also calculated by solving for a in equation (8.2). For this case the computed value was 0.558 ± 0.036 m s^{-2}.

The number of sigmas between the measured value and the calculated value is computed and in using the largest of the two values for σ_a the number of sigmas between the measured and calculated values is 1.2. This is slightly above the desired threshold of 1σ but within the range of this simple experiment.

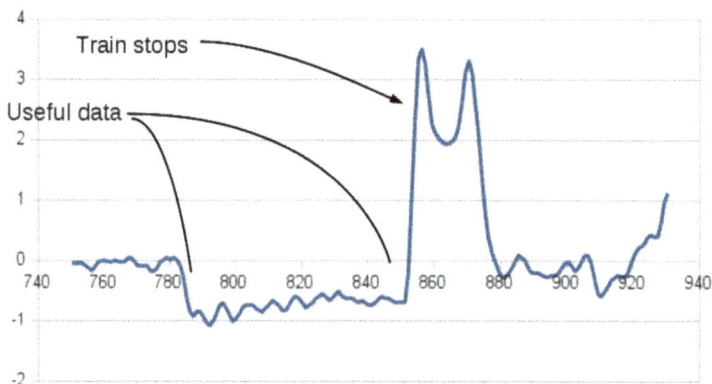

Figure 8.6. A portion of the run starting at $x = 750$ of figure 8.5.

8.5.2 Experiments 2, 3 and 4

The analysis of the other experiments follows the same protocol. The differences being that experiment 2 will use a different value for m_A and experiments 3 and 4 will use their respective equations for the computation. The steps for analysis, however, remain the same.

8.6 Possible problems

There are a few things that may cause problems in this experiment.

1. The train wheels can be off the tracks. The gondola has eight small wheels and all of them need to be on the rails. If there is a clacking sound as the train moves then one of the wheels is probably hitting the railroad ties.
2. The train must be within a usable speed range. If the train is moving too fast then only a few data points will be collected and the smart phone sensor may not be quick enough to provide a proper reading. If the train is too slow then friction will play a larger role. In this case the length of the run was about two-thirds of a meter in 0.75 s.
3. The accelerometer app needs to have a sufficient sampling rate. In this case the sampling rate was 60 Hz which provided 50 sampling points in the 0.75 s.
4. The correct axis must be used. The smart phone measures acceleration in three axes and the correct one must be used.
5. There is too much friction. If the acceleration is not sufficiently constant then there is probably some friction in the system. If the wheels of the train are on the rail then the other source of friction is the string going over the edge of the board. The binder clip is used to reduce this friction and the string should stay centered on the clip during the run.
6. The phone must be aligned. It is imperative that the phone be aligned with the direction of motion. If the phone is resting on the gondola slightly rotated about the vertical axis then the linear motion will be recorded along two of the phone's axes. This will complicate the analysis unnecessarily.

Chapter 9

Lab 5: conservation of energy

9.1 Educational goals

This lab is designed to study the conservation of energy in a system that has no (or very little) energy loss.

9.2 Materials

The materials needed for this lab are:
- a gondola train car and track (see chapter 4),
- a meter stick,
- orientation app for a smart phone,
- accelerometer logger app for a smart phone,
- video camera or burst camera,
- video analysis software, and
- a spreadsheet.

9.3 Theory

For a frictionless system total energy is conserved. Consider the system shown in figure 9.1 which shows an object (a train car) moving along an inclined plain. In this experiment there will be excessive track beyond the boundaries of the experiment

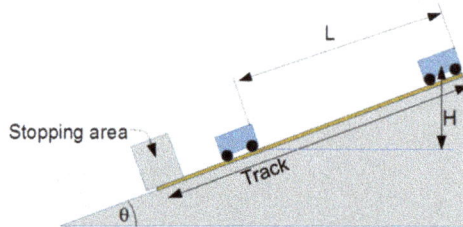

Figure 9.1. The schematic for the conservation of energy experiment.

and so the $y = 0$ plane is the horizontal line from the lower car. The distance to the upper location of the car is H which is measured from this horizontal line. The car will move a distance L during the experiment along an inclined plane at an angle of θ from the horizontal.

Initially, the car is at rest at the upper position and therefore the total energy is just the potential energy, U,

$$E = U = mgH, \tag{9.1}$$

where m is the mass of the gondola train car and the smart phone that it will carry.

The train is allowed to roll freely down the track for a distance L. There is track beyond this location that will be used when bringing the train and phone to a safe stop. As the train moves down the track it will have both kinetic and potential energy,

$$E = K + U = \frac{1}{2}mv^2 + mgh, \tag{9.2}$$

where K is the kinetic energy, v is the velocity of the train parallel to the plane and h is the height of at the location of the train.

If energy is conserved then equations (9.1) and (9.2) should be equivalent. Thus, it is possible to write

$$mgH = \frac{1}{2}mv^2 + mgh, \tag{9.3}$$

and as seen the mass term cancels.

The time starts when the train starts moving and so when the train is at a height H the time is considered to be $t = 0$. It will be possible to measure L, the time of the run t, the angle θ and the acceleration during the run, a_e, where the subscript 'e' indicates that this is the experimental measure of acceleration.

The calculated acceleration, a_c, requires further consideration of the measurable quantities of the experiment. The terms v and h are not easily measured and so it is prudent to express this equation in terms of the variables that the smart phone can measure. Using $h = (L - l) \sin \theta$ and $v = at$ the conservation of energy can be rewritten as

$$gH = \frac{1}{2}a^2t^2 + g(L - l) \sin \theta, \tag{9.4}$$

where l is the distance that the car traveled. At the end of the run $L = l$ and it is easy to solve for a_e to obtain

$$a_e = \frac{\sqrt{2gH}}{t}. \tag{9.5}$$

This describes the calculated acceleration a_c which can be compared to the experimental acceleration. If energy is conserved then $a_c = a_e$ within the precision of the measurements as signified by being less than 1σ.

9.4 Procedure

The train track is set on a flat inclined plane and the angle of this incline is measured using the orientation app. If the app measures only to an accuracy of $1°$ then the angle is represented as $\theta \pm \sigma_\theta$ where $\sigma_\theta = 1°$. Even for small angles this train will move quickly which will limit the number of time samples taken during the run. So, it is more effective to use long runs of track and small angles.

The track is tacked to a piece of wood (see chapter 4) and the beginning and ending locations of the run should be marked on the wood to define length L. It is possible to include the uncertainty σ_L but it is expected that this will be very small when compared to the other uncertainties that are encountered and thus can be neglected. There is excess track beyond the experiment length L that is used when slowing the train to a stop. The train is also carrying the smart phone and so stopping the train safely is important to prevent damage to the phone.

A video or burst camera is positioned to record the trial. Best practice would be to place the phone perpendicular to the plane of the wood to minimize optical and perspective distortions. However, this is not too critical since the sole purpose of the recording is to obtain the time that it takes for the train to move the distance L.

The gondola is placed at the upper mark and the smart phone is placed on it. The app that records acceleration is turned on. In this lab there is no need to synchronize the timing of the video sequence with the timing of the accelerometer readings. If, however, the reader wishes to modify the experiment making this synchronization necessary, then the train should be moved quickly a small distance by hand thus causing a spike in the recorded acceleration. This can then later be synchronized with the visible movement in the recording that caused the spike.

The train is released and allowed to run the distance L unimpeded. The train is brought to a safe stop and the app is turned off. The recorded data are then sent to a computer for further analysis.

It is required that a video be recorded during the run but the smart phone is occupied because it is recording acceleration. Other options for recording video are a digital camera, a second smart phone, or a camera that is embedded in a laptop.

9.5 Analysis

The duration of the run is extracted from the video using video analysis software (see section 1.3.2) and is represented as $t \pm \sigma_t$. As the train moves faster it will be become more blurred which will cause uncertainty in the measurement of t. Generally, burst camera apps are better at capturing faster motion that video cameras. Then equation (9.5) can be used to calculate the acceleration $a_c \pm \sigma_{a_c}$.

The experimental acceleration is obtained from the accelerometer data. This process will follow that of section 8.5.1. There will be a lot of data collected before and after the run that are to be discarded. Identification of the pertinent data starts with the location of the large spike that is caused by the phone coming to a stop. The useful data are just before that and should be a (mostly) flat sequence of readings. Negative values for acceleration merely mean that the phone was facing the wrong way. The experimental value of acceleration is the average and standard deviation of these data points, $a_e \pm \sigma_{a_e}$. Again section 8.5.1 has a more detailed explanation of where to find the data within the recording.

If the energy is conserved then the value of $|a_e - a_c|$ should be less than one standard deviation (the largest of σ_{a_e} or σ_{a_c}).

There are a few optional experiments that can also be explored.

1. Add more mass to the train (by placing weights in the car). This should not affect the time of the run since the mass canceled out of the equations.
2. Run the experiment with different values of θ to show that $a_c \propto \sqrt{H}$.
3. Compute the energy at a distance of $L/2$ and use equation (9.4).

9.6 Possible problems

There are a few aspects that can cause problems in this experiment.

- If θ is too large then there will be an insufficient number of samples from the accelerometer.
- If θ is too small then friction will be significant in the calculations.
- If the track is too short then it will be difficult to obtain a sufficient number of samples.

There are actually a few items that are not problematic although they may seem so at first.

- The board the track is on is not flat. In a system without friction the path that the object takes does not affect the velocity at the end of the run. Small imperfections in flatness are not too harmful.
- Usually it is important to have the camera perpendicular to the action to minimize distortion. In this case the camera is only used to obtain the start and end times which are when the train passes the markings on the board.

Chapter 10

Lab 6: loss of energy

10.1 Educational purpose

This lab is designed to study the conservation of energy in a system that instantaneously loses a fraction of its kinetic energy.

10.2 Materials

This lab will need the following items:
- a basketball,
- a smart phone with video capabilities,
- a meter stick,
- sidewalk chalk,
- a spreadsheet, and
- video analysis software.

10.3 Theory

The conservation of energy means that the total energy is constant throughout the motion of an object. If there are no friction or losses then the total energy is the sum of the potential energies plus the kinetic energy. For a system that has losses then this loss must also be taken into consideration.

In this experiment a basketball is thrown high in the air and allowed to bounced several times as shown in figure 10.1. This toss will impart a nontrivial horizontal velocity as well. Each time the ball bounces there is a significant loss of energy due to the deformation of the ball. The loss due to air friction during flight will be considered too trivial to matter. This experiment will measure the times of the bounces and the horizontal distance between the bounces. Thus the data collection begins when the ball first touches the ground at t_1.

During the flight between the bounces energy is conserved without loss and so $E = U + K$ where U is the potential energy and K is the kinetic energy. Since there are

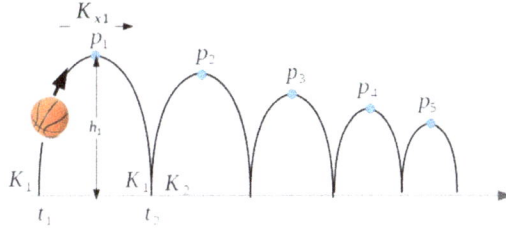

Figure 10.1. The basketball is allowed to bounce several times and the peak heights are noticeably smaller after each bounce.

no losses during this flight the kinetic energy just after t_1 and just before t_2 are the same. The total energy at these times is

$$E = K_1. \tag{10.1}$$

At the peak there is potential energy and still some kinetic energy since $v_x \neq 0$. Thus, the total energy at the peak is

$$E_1 = U_1 + K_1 = mgh_1 + \frac{1}{2}mv_{x1}^2. \tag{10.2}$$

Furthermore, the time of flight from t_1 to the peak, t_{1p}, is half of the total flight time,

$$t_{1p} = \frac{t_p - t_1}{2}. \tag{10.3}$$

Since acceleration is constant $h = \frac{1}{2}gt_{1p}^2$ and therefore

$$U_1 = mgh_1 = \frac{1}{2}mg^2t_{1p}^2. \tag{10.4}$$

The horizontal distance that the basketball travels between times t_1 and t_2 will also be measured and is assumed to be constant between the bounces.
Thus,

$$v_{x1} = \frac{d_{12}}{t_{12}}, \tag{10.5}$$

where d_{12} is the distance between the first and second bounce. Thus, it is possible to calculate the energy of the ball from equation (10.2) which is also the value of the kinetic energy K_1. Since v_{x1} is known the vertical velocity can also be determined by solving for v_{y1}:

$$K_1 = \frac{1}{2}mv_{x1}^2 + \frac{1}{2}mv_{y1}^2. \tag{10.6}$$

At the bounce a percentage of the kinetic energy is lost which is described as

$$K_2 = \alpha K_1, \tag{10.7}$$

where α is a value between 0 and 1 that will be determined in this experiment. If the same percentage of kinetic energy is lost at each bounce then α will be a constant:

$$K_{i+1} = \alpha K_i, \quad \forall i. \tag{10.8}$$

The magnitude of the loss of energy, L_i, for bounce i is written as

$$K_1 = K_{i+1} + L_i. \tag{10.9}$$

If the percentage of loss is the same for each bounce then α will be a constant and L_i will be exponentially decaying with respect to the bounce index i. This will be confirmed by creating two plots and applying a curve fit to these curves. The first will plot α_i versus i and the second will plot L_i versus i.

10.4 Procedure

The procedure for this experiment is quite simple. This experiment will require a hard, flat, horizontal surface and plenty of headroom. If possible the surface is marked with chalk at regular intervals such as 50 cm as shown in figure 10.2. The alternative to marking on the surface would be to position a meter stick as shown.

The basketball is thrown up in the air and allowed to bounce with a horizontal motion that is parallel to the run of chalk marks or meter stick. This motion is recorded as a video or series of burst camera images. It is important that the smart phone be directly facing the vertical plan of motion so as to minimize lateral skew in the images. Video analysis or the burst camera time stamps are used to extract the time of each bounce. It will not be possible to pinpoint the exact time of the bounce since the ball will spend a finite amount of time on the ground during the bounce. Thus each time measurement will have an uncertainty, $t_i \pm \sigma_{t_i}$ where i is the bounce number.

The horizontal distance between each bounce is measured by the number of pixels between bounce points. To convert this to meters the number of pixels is multiplied by ρ which is the ratio of the meter (rule or chalk marking) to the number of pixels that 1 m consumes in the image.

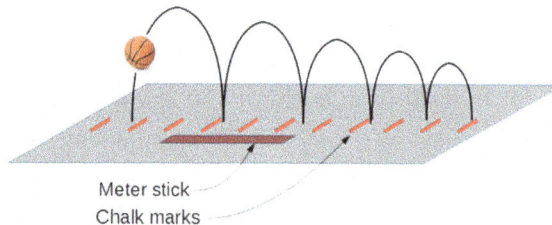

Meter stick
Chalk marks

Figure 10.2. The horizontal motion of the ball follows the chalk marks or the meter stick.

10.5 Analysis

For each bounce several parameters are computed.

- The potential energy at each peak is calculated using equation (10.4). Since there is an uncertainty in the measurement of time there will also be an uncertainty in the calculation of the potential energy which will be represented as $U_i \pm \sigma_{U_i}$.
- The horizontal velocities for each segment are calculated from the distance between the bounces and times between the bounces. While the horizontal velocity is constant between the bounces, the action occurring during the bounce subtracts energy from the ball. Thus, the velocity after each bounce i is represented by $v_i \pm \sigma_{v_i}$.
- The kinetic energy at the top of each peak is calculated using $K_i = \frac{1}{2}mv_{xi}^2$ and expressed with its uncertainty as $K_i \pm \sigma_{K_i}$.
- The total energy at each peak is calculated using equation (10.2). Including uncertainty calculations the total energy of the ball at each peak is represented by $E_i \pm \sigma_{E_i}$.
- This calculation also provides a numerical value for the kinetic energy just above the ground. From this the fraction of the retained energy, $\alpha_i \pm \sigma_{\alpha_i}$ and the magnitude of the energy loss, $L_i\sigma_{L_i}$, for each bounce is calculated.

The values of α_i and L_i are each plotted versus the bounce number i. An appropriate curve fit (using Excel Trendline) and the computed R^2 values from these fits will indicate if the hypothesis of a constant α value is upheld.

Many of the calculations have m, the mass of the basketball. This is actually not an important value as α can be computed without knowing m. The value of L does require m but it is a constant and so the functional form of the L_i will not be affected by the value of m. So, the experiment can be performed without this value or the value could be obtained from measuring the weight of the ball or through a web search.

10.6 Possible problems

There are a few aspects that could cause problems in this lab.

- The basketball will have a horizontal velocity and this trajectory should be parallel to the chalk markings or meter stick. If it is not then the distances between the bounces will be more difficult to estimate.
- The ground should be level. If it is slanted then there will be a transfer of vertical kinetic energy with the horizontal kinetic energy which is not currently included in the calculations.
- The camera should face perpendicular to the vertical plane of motion. If it is at an angle then the horizontal distances will be more difficult to estimate.

Chapter 11

Lab 7: inelastic collisions

11.1 Educational goal

This lab is designed to explore the loss of energy in inelastic collisions.

11.2 Materials

The materials needed for this lab are:
- two gondola cars,
- train track,
- Velcro,
- a burst camera or video camera, and
- a meter stick or a calibration chart.

11.3 Theory

In an *elastic collision* the momentum is conserved but the kinetic energy is not. This lab will explore the loss of the kinetic energy under varying conditions. Consider a case in which there are two masses m_1 and m_2 with respective velocities v_1 and v_2. After a perfectly inelastic collision the masses are fused together creating a single mass m_f which is moving at a velocity v_f. The conservation of momentum is

$$m_1 v_1 + m_2 v_2 = m_f v_f. \tag{11.1}$$

The kinetic energy, however, is not conserved:

$$K_1 + K_2 > K_f, \tag{11.2}$$

where $K_1 = \frac{1}{2}m_1 v_1^2$, $K_2 = \frac{1}{2}m_2 v_2^2$ and $K_f = \frac{1}{2}m_f v_f^2$. This loss can be expressed as a ratio of the total kinetic energy after the collision to the total kinetic energy before the collision:

$$f = \frac{K_f}{K_1 + K_2} = \frac{\frac{1}{2}(m_1 + m_2)v_f^2}{\frac{1}{2}m_1 v_1^2 + \frac{1}{2}m_2 v_2^2} = \frac{(m_1 + m_2)v_f^2}{m_1 v_1^2 + m_2 v_2^2}. \tag{11.3}$$

Analysis by Mungan [1] reduced this equation by applying four steps.

1. Multiply the numerator and denominator by $\dfrac{1}{v_2^2}$.

2. Define $r \equiv \dfrac{v_1}{v_2}$.

3. Multiply the numerator and denominator by $\dfrac{1}{m_2}$.

4. Define $\mu \equiv \dfrac{m_1}{m_2}$.

The ratio of kinetic energy after the collision to the kinetic energy before the collision becomes

$$f = \frac{(\mu r + 1)^2}{(\mu + 1)(\mu r^2 + 1)}. \tag{11.4}$$

Figure 11.1 shows the values of f versus r for three different values of μ. Consider the case of $\mu = 1$. At $r = 1$ the ratio is $f = 1$. Recall that r is the ratio of velocities and so when $r = 1$ the two objects are moving in the same direction at the same speed. This is really not a collision and so no kinetic energy is lost, thus $f = 1$.

At $r = -1$ the speeds of the two masses are equal but in opposite directions. The masses are also the same since this is the case of $\mu = 1$. Thus, when they collide they will stick together and come to a complete stop. In this case there is a total loss of kinetic energy and so $f = 0$.

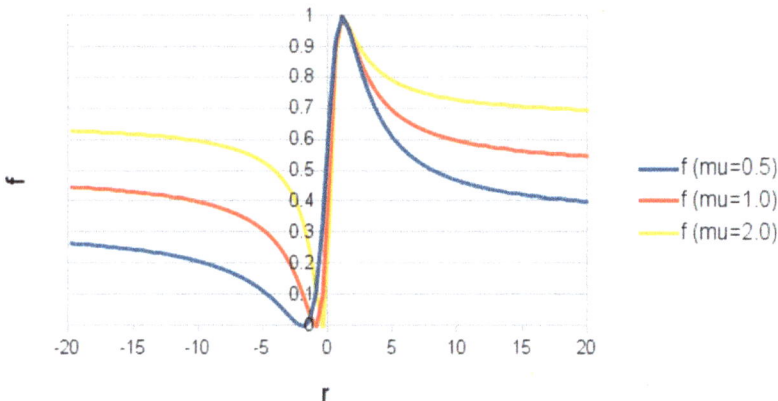

Figure 11.1. Plots of f versus r for three different values of μ.

11-2

The case of $r = 0$ indicates that $v_1 = 0$ which means that m_2 is moving and m_1 is stationary. In this case $f = 0.5$ which indicates that half of the kinetic energy is lost. In the case of $r = \infty$ or $r = -\infty$ just the opposite is true. m_1 is moving and m_2 is stationary but since the masses are the same the loss of kinetic energy is also half.

The cases of $\mu = 0.5$ and $\mu = 2.0$ are also shown, which are cases in which one mass is twice as large as the other. The case of $\mu = 0.5$ and $r = \infty$ is the same situation as $\mu = 2$ and $r = 0$. In these cases the larger mass is moving and so the loss of kinetic energy is less than half.

This lab will collect data from different trials and compare the values to those on the graph.

11.4 Procedure

The experiment is shown in figure 11.2 where there are two train gondola cars with extra weights. Each car has a strip of Velcro on the end so that they will stick together upon collision. The train cars do come with couplers but they easily break off and so Velcro tends to work better. Behind the experiment is a calibration chart which has marks at regular intervals. A meter stick could be used but often the numbers are hard to read at video resolution.

The mass of each gondola with its weights is measured. This measurement will have a standard uncertainty and thus the mass of the first car and its weight is $m_1 \pm \sigma_{m_1}$. In the first set of trials the masses of the two cars should be nearly equal. The burst or video camera is positioned such that it is perpendicular to the train track.

In the first trial the cars are gently pushed towards each other at approximately the same speed. This experiment does have some delicacy in it and therefore it is wise to try a few practice runs first at slow speeds gently increasing the speed with each run until a suitable speed is assured. Both cars are pushed towards each other and collide.

The speed of the cars is extracted from the burst images. Since there is some friction in the system it is best to obtain the velocities of the two cars just before the collision and just after the collision. This is achieved by obtaining two video frames just before the collision. The distance that the car traveled is extracted from the calibration chart and the time of the travel is extracted from the frames' time stamps. This provides the velocity before the collision $v_i \pm \sigma_{p_i}$. The process is repeated to compute the velocity just after the collision $v_f \pm \sigma_{v_f}$.

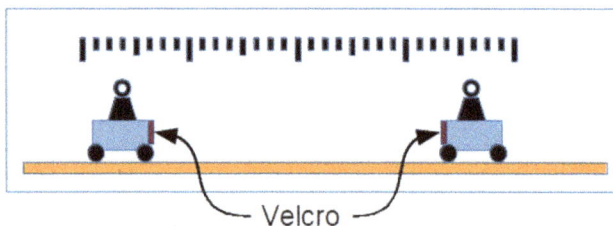

Figure 11.2. The schematic for this experiment. Two gondolas with weights are placed on the track and a calibration board is placed within the camera's field of view.

This experiment is repeated several times with different speed conditions. In some cases the speeds should be similar, in others they should be significantly different, and in a few one of the cars should be stopped. This will provide a set of different r values. In the analysis the values of $f \pm \sigma_f$ will be determined and compared to the theoretical values.

An option is to repeat this experiment such that one mass is close to twice as large as the other thus providing trials with different values of μ.

11.5 Analysis

The analysis for one set of trials with a constant μ begins with the calculation of μ. Then a plot similar to those shown in figure 11.1 is created for this μ value. This is called the K-loss graph.

Several trials are performed and for each trial the value of r is calculated from the ratio of the velocities. With the masses and the velocities the ratio of the kinetic energies is calculated by

$$f = \frac{K_{\mathrm{f}}}{K_1 + K_2}. \tag{11.5}$$

Since there are uncertainties in the velocities (due to the uncertainties in the distances d) the computed value is represented by $f \pm \sigma_f$.

On the K-loss graph each value of $f \pm \sigma_f$ is also plotted. For a specific value of r there is an associated $f \pm \sigma_f$. A marker is added to the plot at location (r, f) and vertical error bars are used to show the standard deviation. An example is shown for one data point in figure 11.3. In this example, $r = -6$, $f = 0.39$ and $\sigma_f = 0.18$. The fact that the function (curve) passes through the error bar means that the calculated value (curve) and the experiment value (plotted point) are within one sigma and contribute to confirming the theory.

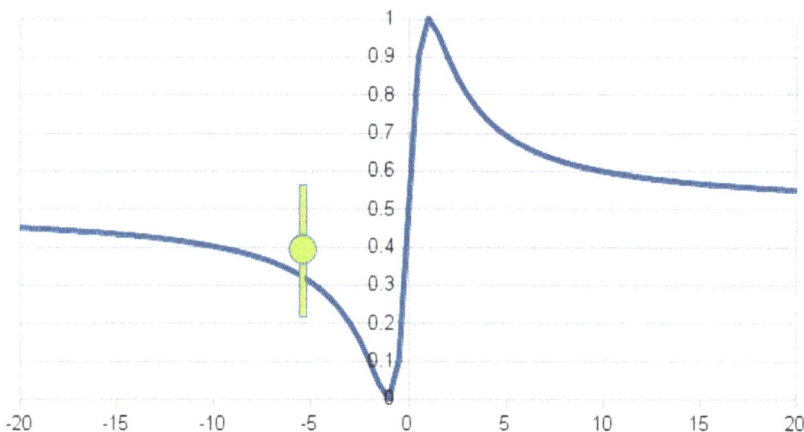

Figure 11.3. Adding one data point to the graph of the ratios of kinetic energies.

Points and error bars for all the data collected for this value of μ are likewise added to the plot. The conclusion is then based on the number of error bars that intersect with the curve. If experiments with different values of μ were performed then a different graph is created with the appropriate curve.

11.6 Possible problems

There are some problems that can easily occur in this experiment.
- If the weights are too small then there is very little energy at the beginning of the experiment and thus friction and computational uncertainties will play a significant role.
- If the cars are moving too fast then it will be difficult to obtain a meaningful value for d.
- If the cars are moving too slowly then other losses of kinetic energy will play a role but are not accounted for in the theory.
- If the weights in the cars shift during the collision then that motion will absorb energy and skew the results.
- If all data points are lower than the curve then there is an absorption of energy other than the collision.

Bibliography

[1] Mungan C E 2013 Mechanical energy changes in perfectly inelastic collisions *Phys. Teacher* **51** 229–30

Chapter 12

Lab 8: angular acceleration

12.1 Educational goal

This lab extracts a value of angular acceleration from a spinning object.

12.2 Materials

The materials needed for this lab are:
- an object that spins about a vertical axis (office chair, lazy Susan, etc),
- an app to log orientation versus time, and
- a spreadsheet.

12.3 Theory

The theory for this lab is straightforward. The angular velocity is defined as

$$\omega(t) = \frac{d\theta}{dt}, \tag{12.1}$$

where θ is the angular displacement of the object as a function of time. Likewise, the angular acceleration is defined as

$$\alpha(t) = \frac{d\omega}{dt}. \tag{12.2}$$

In this lab the motion of an object will be estimated with a polynomial,

$$\theta(t) = at^2 + bt + c, \tag{12.3}$$

where a, b and c are constants that are determined by curve fitting software. Once these constants are known the functions for angular velocity and angular acceleration are easily determined:

$$\omega(t) = 2at + b \tag{12.4}$$

and

$$\alpha(t) = 2a. \tag{12.5}$$

12.4 Procedure

Like the theory, the procedure for this lab is straightforward. A smart phone is placed on the object that will spin, such as an office chair. The phone is placed on the seat of the chair such that the phone is horizontally flat and located at the center of the axis of rotation.

The app that logs orientation angles is turned on and the chair is spun so that it rotates without moving laterally. The chair is allowed to come to a stop and the app is turned off. The collected data are sent to a spreadsheet for further analysis.

The sensor will collect orientation data for all three axes of the smart phone. The only column of data that is important is the one associated with the vertical axis. Example data are shown in figure 12.1.

12.5 Analysis

There are three items within the raw data that are important to recognize. The first is that it has a sawtooth appearance. The angle increases until it reaches 360° which is is equivalent to 0°. Thus, the collected data have a sharp vertical line each time the orientation advances from 359° to 0°. As seen in these data this occurred four times because the chair spun for four revolutions before coming to a stop. If the chair is spun in the opposite direction then the sawtooth pattern will be reversed.

The second item is that there are data at the beginning of the trial that need to be removed. The only data that are important are those when the chair is spinning freely without anyone touching it. Thus the first few seconds need to be removed. As seen here the phone is at rest (horizontal portion) and then spun up to an initial

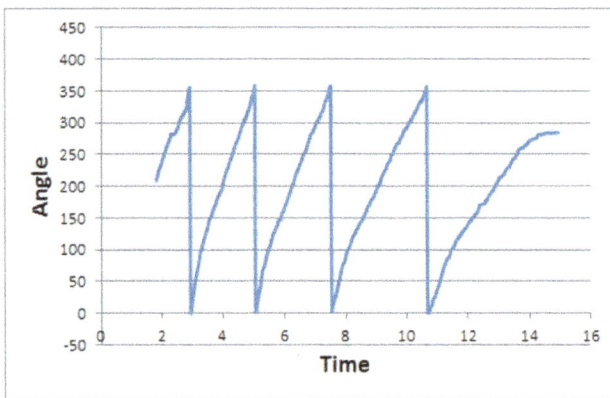

Figure 12.1. The raw data collected from the orientation sensor.

velocity (jagged increasing line). Both of these regions should be neglected in the rest of the analysis.

The third item is that at the end of the trial there is another flat region in which the chair comes to a stop. At the stopping point static friction will become significant changing the acceleration. Thus, the final part of the data collection should also be discarded.

To complete the numerical analysis the sawtooth nature of the data needs to be removed. Each time the chair makes a revolution the displacement is 360° which is absent from the collected data. Thus, after each revolution 360° needs to be added to the data. In this case all data after the first revolution, $t > 3$, have 360° added to them, all data after $t > 5$ have 720° added to them and so on. These calculations are performed in the spreadsheet and the result is shown in figure 12.2. This charts the angular displacement with respect to time without the wrap around effect seen in figure 12.1.

The next step is to estimate this curve with a polynomial using the Trendline function in Microsoft Excel (see section 2.4). Trendline estimates the function that fits the example data as

$$\theta = -5.35t^2 + 191t + 175, \qquad (12.6)$$

which defines the values a, b and c in equation (12.3). Once these are known then the functions $\omega(t)$ and $\alpha(t)$ can be computed using equations (12.4) and (12.5).

Finally, if the same chair is used in multiple trials then it is expected that the frictional forces that are causing the chair to slow down are unchanged. Therefore, the value for the angular acceleration should be the same for multiple trials. With these multiple trials it is possible to calculate $\alpha \pm \sigma_\alpha$ and in successful experiments $\sigma_\alpha \ll \alpha$.

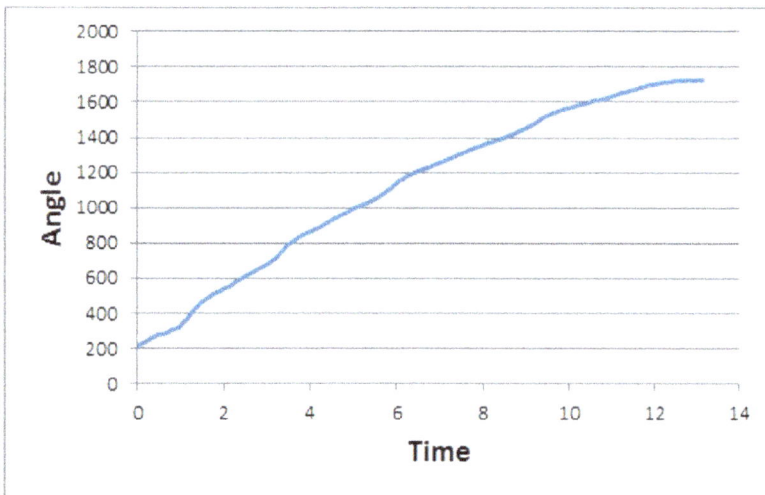

Figure 12.2. The data after the tails and the wrap-around effect are removed.

12.6 Possible problems

There are a few problems that can cause poor experiments.

- If the chair or spinning object does not reach several revolutions before coming to a stop then its friction is too great. In the case of an office chair adding weight to the chair can help increase the number of revolutions. Otherwise, it may be necessary to find a different object.
- If the chair wobbles while spinning or has lateral motion then the results will be poor.
- If the chair is old then it is possible that it will not spin smoothly and angular acceleration will not be constant.

If an object to be used as the spinner is not readily available then one will need to be constructed. This can be a simple piece of wood with a nub on one face of the wood located at the center of mass which can be as simple as a protruding nail head. This will rest on the table as the board spins. This construction will have many revolutions but will also be more difficult to use since it must be balanced throughout the experiment.

Chapter 13

Lab 9: efficiency of momentum transfer

13.1 Educational goal

The goal of this lab is to understand the transfer of angular momentum to linear momentum.

13.2 Materials

The materials needed for this lab are:
- scales,
- four similar toy cars with metal axles (e.g. Hot Wheels cars),
- a meter stick,
- a burst camera app, and
- track for the toy cars (optional).

13.3 Theory

The relationship between rotational motion and linear motion is expressed as

$$v = \omega r, \tag{13.1}$$

where v is the linear velocity, ω is the angular velocity and r is the radius from the mass to the center of rotation. Consider the system shown in figure 13.1 which shows a mass m initially at rest against a lever positioned at points $\overline{PA}$. The lever is then rotated about point P through an angle θ with a constant angular velocity ω. When the lever reaches the position $\overline{PB}$ the swinging end strikes a block that stops the lever. The mass, however, will continue up in a linear motion in the direction shown by d.

During the linear motion there is significant friction and the mass will come to a stop at a distance d from where the lever stopped. Just as the mass leaves the lever it has an initial linear velocity of v and this is related to the angular velocity by equation (13.1). Measuring this velocity is difficult but it is related to d by

$$v^2 = 2ad, \tag{13.2}$$

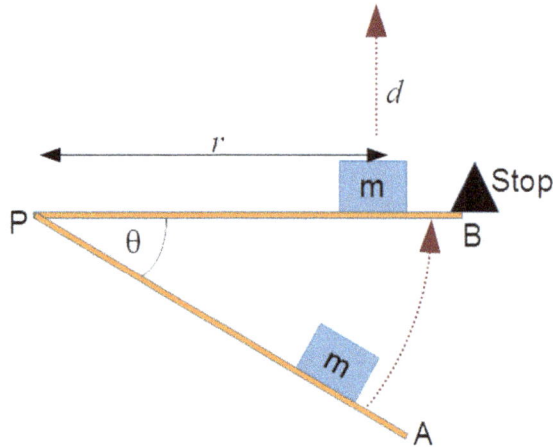

Figure 13.1. A schematic of this lab. A mass is moved by a rotating stick and then allowed to move in a linear manner.

where a is the negative-valued acceleration due to friction. The distance d can easily be measured but the acceleration is computed by measuring the time t_2, which is the duration of the linear motion, by

$$a = \frac{2d}{t_2^2}. \tag{13.3}$$

Thus, it is possible to determine the left side of equation (13.1) with an uncertainty and this is represented as $v \pm \sigma_v$.

The distance r can also be easily measured but the angular velocity is determined from knowing the angle θ and the time t_1 that the lever was in motion. The angular velocity is then computed by

$$\omega = \frac{\theta}{t_1}, \tag{13.4}$$

and is represented as $(\omega r) \pm \sigma_{(\omega r)}$. Thus, it is possible to determine the right side of equation (13.1).

Support for equation (13.1) is given if $v = \omega r$ within one standard deviation.

13.4 Procedure

Running several trials with the same ω will be difficult and so this experiment will run four experiments in a single trial. Figure 13.2 shows the experiment which is based on a system shown by Haugland [1]. Four slots just big enough for the cars are cut into a block of wood. A nub is placed on the bottom of the wood to provide a pivot point. This can be as simple as a nail that is poking through the bottom of the wood.

Four identical cars are placed in the slots. The toy cars should be identical because the coefficient of friction varies greatly between different models. These cars should also have metal axles instead of plastic ones for the same reason.

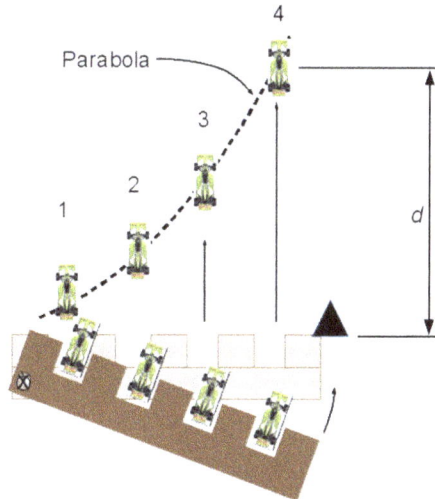

Figure 13.2. A block of wood with slots is used as the lever and four identical cars are the masses.

The video camera is placed overhead to capture the motion of the lever. If this same video camera is not able to also capture the linear motion of the cars then a second camera is used to capture that motion. The purpose of these cameras is to capture the amount of time that the lever is in motion and the amount of time that the cars are in linear motion.

Before the cars are set into place the distances r_i are measured which are the distances from the center of each slot to the pivot point. This should be a radial distance and so this measurement should be parallel to the wood when it is stopped as shown in figure 13.1. The masses of the cars are measured as well.

The block is positioned at a set angle θ from where it will stop. With the cameras on, the block is rotated through an angle θ. It is important to move the wood with a constant ω. The block of wood suddenly stops and the cars begin their linear travel. The cars come to rest and the distances d_i from the stopped wood to the stopped cars are measured. If a flat smooth surface is not available then an option would be to allow the cars to run along car track. It is also important that the cars run in a straight line. Cars that veer significantly will create more friction when running along toy car track.

Since the distances d_i should be proportional to r_i^2 the pattern that the cars make when stopped should be parabolic as shown in figure 13.2. If this did not occur then the experiment needs to be repeated.

The distances d_i and r_i are measured directly. The angle θ is established at the beginning of the experiment and the time of the swing, t_1, is measure from the camera. Each car will have its own stopping time t_2 and these are also measured from the camera.

13.5 Analysis

There are two parts for the analysis which are to compute both sides of equation (13.1). The first part is to compute the value v_i from equation (13.2) using equation (13.3)

13-3

to compute the acceleration a. Identical cars should have nearly identical values of a. The second part is to compute the right side of the equation (ωr). The value ω is computed from equation (13.4) and is the same for all cars. The value r_i is measured directly.

A single experiment will have four trials. There are two reasons for this. The first is that it can indicate the trial was successful if the cars stop in a parabolic pattern as shown in figure 13.2. If the stopping pattern of the cars is not parabolic then the experiment needs to be attempted again. The second is that it provides four different values of v and four different values of ωr. Thus the standard deviations can be computed to provide $v \pm \sigma v$ and $(\omega r) \pm \sigma(\omega r)$.

The experiment supports equation (13.1) if $v = \omega r$ within one standard deviation.

13.6 Possible problems

There are few aspects that can interfere with this experiment.

- Identical cars must be used. Different model cars have vastly different coefficients of friction.
- Cars should run straight even if track is being used. Usually, a new car will run fairly straight, but after handling the thin wire axles tend to bend upwards thus causing the car to veer off to the side. Even if the car is running along track this alteration to the car will cause it to ride the rail of the track increasing its coefficient of friction.
- The cars should have metal axles. Cars with plastic axles have higher friction and are less uniform.
- The slots in the wood can cause friction as the car leaves. The slots should not be too deep. In figure 13.2 the slots are about half of the length of the cars. They should not be deeper than this.
- The motion of the lever should be as uniform as possible to create a constant ω over the angle θ.

In this experiment there can be clear evidence that the trial did not work well. Since d is proportional to r^2 the stopping locations of the cars should provide a parabolic pattern. If this is not the case then the trial needs to be discarded and attempted again.

Bibliography

[1] Haugland O A 2013 Car stopping distance on a tabletop *Phys. Teacher* **51** 268

Chapter 14

Lab 10: angular momentum

14.1 Educational goal

The goal of this experiment is to show that angular momentum is conserved.

14.2 Materials

The materials needed for this experiment are:
- a wire clothes hanger,
- needlenose pliers and wire snippers,
- a meter stick,
- duct tape,
- fishing line or strong string that does not stretch,
- a metal keyring or paper clip,
- two rubber bands,
- six metal nuts,
- scales, and
- a video or burst camera.

14.3 Theory

In a system spinning without friction angular momentum is conserved. The expression for angular momentum that is useful here is

$$L = I\omega, \tag{14.1}$$

where I is the moment of inertia and ω is the angular velocity.

Figure 14.1 shows a system that consists of a rod of mass m_2 and length l rotating in a horizontal plane about an vertical axis. There are also two weights each of mass m_1 and each a distance r from the axis. In this system the moment of inertia is

$$I = \frac{1}{12}m_2 l^2 + 2(m_1 r^2). \tag{14.2}$$

doi:10.1088/978-1-6270-5628-1ch14

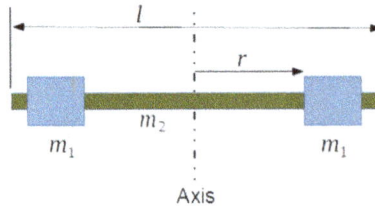

Figure 14.1. This system consists of a stick of mass m_2 rotating about an axis with two equal masses of m_1 a distance r from the axis.

There are two phases in this experiment. In the first phase $r = r_A$ and in the second phase $r = r_B$ where $r_A > r_B$. Thus, $I_A > I_B$ and therefore in order for angular momentum to be conserved, $\omega_A < \omega_B$. Basically, when the radius of the weights is decreased then the angular velocity increases.

In this experiment the system will spin at ω_A and then at ω_B. The angular momentum for both cases will be determined along with their uncertainties, $L_A \pm \sigma_{L_A}$ and $L_B \pm \sigma_{L_B}$. The conservation of angular momentum is supported if $L_A = L_B$ within one standard deviation.

14.4 Procedure

This lab requires some dedicated effort towards its construction. Figure 14.2(*a*) shows a wire hanger in which the lower wire has been cut in the middle. A mass m_1 is constructed by taping together three or four metal nuts so that they will slide onto the metal wire. A rubber band is attached to a nut at one end and a long string or fishing line is attached to the other. The set is then wrapped into duct tape to make a single mass. This mass is measured on the scales. A second mass is created in a similar fashion. The mass m_2 of the hanger is also measured.

The nut assemblies are slid on to the wire and the rubber bands are attached to the ends of the hanger. This should be done such that the length of the unstretched rubber bands are the same. The string side of the nut assemblies face towards the center of the hanger wire.

The ends of the cut wire are curled with the pliers as shown in figure 14.2(*b*). The keyring is placed on one of these wires and the curled wire ends are hooked onto each with one part of the keyring in between them as shown in figure 14.2(*c*). The string from the assemblies is threaded through the keyring and allowed to hang down. The goal is to create a system such that when the string is pulled downwards the weights move towards the center of the system in an equal manner. When the tension of the string is relaxed the rubber bands will bring them back to their original positions.

Now that the apparatus is constructed it is hung by a string at the top of the hanger so that it can hang and spin freely. The video camera is placed such that it can capture the spinning motion of the hanger. The goal is to be able to count the number of revolutions and to obtain the amount of time required for the revolutions using video analysis software.

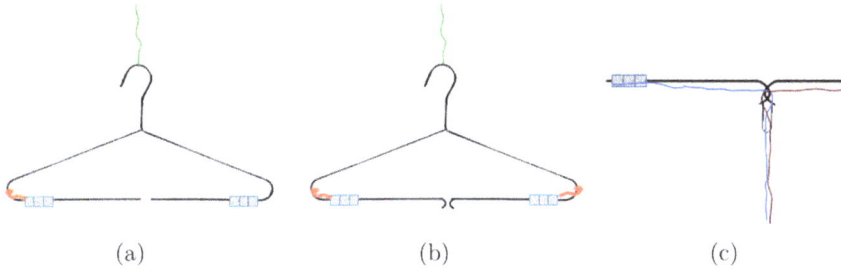

Figure 14.2. (*a*) The system is constructed from a wire hanger and two weights. (*b*) The wire is crimped after the weights are put on. (*c*) String is tied to the weights and channeled through the paper clip such that when the strings are pulled both weights mover inwards.

With the camera on, the hanger is manually spun and allowed to spin for several revolutions. Then the string is pulled downwards causing the weights to change positions and the hanger is allowed to spin for several revolutions. From the video the time required to spin a number of revolutions with weights at r_A is collected. The same types of data are collected with the weights at position r_B. The video also provides the distances r_A and r_B in terms of number of pixels. The ratio of centimeters per pixel can be obtained from the width of the hanger. The width of the hanger can be measured with a meter stick and the number of pixels for that width can be extracted from a video frame. This ratio can convert the pixel measures to centimeters for both r_A and r_B.

14.5 Analysis

From the video it will be possible to count the number of revolutions that occurred in time t_1. This leads to the computation of $\omega_A \pm \sigma_{\omega_A}$ from $\omega = \frac{\theta}{t_1}$ where θ is 2π multiplied by the number of revolutions.

The radius r_A is also determined from the video. The width l (from figure 14.1) of the hanger is measured with the meter stick. This same extent is viewed from a video frame in which the hanger is perpendicular to the camera. The ratio ρ is the length l divided by the number of pixels in the image frame. The number of pixels from the center of rotation to the center of mass m_1 is extracted and multiplied by ρ to determine the distance $r_A \pm \sigma_{r_A}$. From this and the measurement of mass the moment of inertia $I_A \pm \sigma_{I_A}$ can be determined.

The first term in equation (14.2) is related to the angular momentum of the hanger which is not exactly a rod. However, since the mass of the hanger is somewhat evenly distributed with respect to the radial distance the assumption that it behaves like a solid rod will be used. The hook at the top of the hanger does interfere with the idea that the mass is equally distributed versus r but since it is near the center of rotation is effect will be minimal. Furthermore, the hanger does not change during the experiment and as long as the weights of the nuts are dominant this assumption of the moment of inertia for the hanger will not be significant.

The same process is repeated to determine the moment of inertia, $I_2 \pm \sigma_{I_2}$, after the string is pulled. Now it is possible to compute both $L_A \pm \sigma_{L_A}$ and $L_B \pm \sigma_{L_B}$ and if angular momentum is conserved then $L_A = L_B$ within one standard deviation.

14.6 Possible problems

There are a few problems that could occur in this experiment.

- The hanger needs to spin smoothly about a vertical axis. If the system is wobbling then the results will be poor.
- It is possible that when the string is pulled that the nut assembly will want to rotate about a horizontal axis perpendicular to the weight. Basically it will tilt. This is mitigated by ensuring that the string and rubber bands pull from the inside portion of the nuts and not the perimeter.
- If there is friction in the spinning motion then angular momentum is not conserved. Fishing line works best for hanging the system. It is difficult to create a frictionless system and thus it may be necessary to keep the number of revolutions to a small number.
- When the string is pulled it must be held such that r_B does not change.

Chapter 15

Lab 11: torques in equilibrium

15.1 Educational goal

The goal of this lab is to study the forces and torques of a system in equilibrium.

15.2 Materials

The materials needed for this lab are:
- a meter stick,
- string,
- a weight (near the same weight as the meter stick),
- a smart phone camera, and
- image analysis software (such as GIMP).

15.3 Theory

Figure 15.1 shows a system of forces and torques in equilibrium. This system consists of the meter stick suspended from two strings. The left string makes an angle of α to the vertical and the right string makes an angle of β to the vertical. The meter stick makes an angle of θ to the horizontal. The stick has a length l and weight w_1 and at a distance of r there is a second weight w_2. Since this system is in equilibrium the sum of the horizontal forces is zero, the sum of the vertical forces is zero and the sum of the torques is also zero.

The first horizontal force is from the tension of T_1 and it is $T_{1x} = T_1 \sin(\alpha)$ where T_1 is not yet known. Likewise, $T_{2x} = T_2 \sin(\beta)$. These are the only two horizontal forces and therefore

$$T_1 \sin \alpha = T_2 \sin \beta. \tag{15.1}$$

There are four vertical forces, two from the tensions and two from the weights in the opposite direction. Since the sum of these forces is zero,

$$T_1 \cos \alpha + T_2 \cos \beta - w_1 - w_2 = 0. \tag{15.2}$$

doi:10.1088/978-1-6270-5628-1ch15

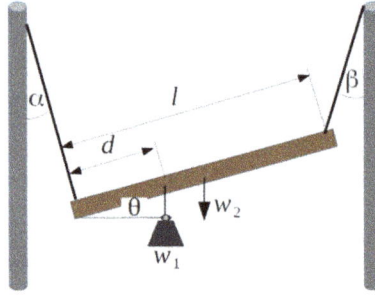

Figure 15.1. A stick suspended at both ends with a weight attached.

Finally, using the left end of the stick as the pivot there are three torques with one from T_2 and the other two in the opposite direction from the weights. Since the torques sum to zero,

$$dw_2 \cos\theta + \frac{l}{2}w_1 \cos\theta = lT_2 \sin\alpha. \tag{15.3}$$

In this lab three variables T_1, T_2 and θ are considered to be unknown and therefore need to be calculated. The calculated value of the angle is represented as θ_c. This angle is also measured experimentally and that measurement is denoted as θ_e. Solving equation (15.1) for T_2 and then substituting that into equation (15.2) yields

$$T_1\left(\cos\alpha + \frac{(\sin\alpha)(\cos\beta)}{\sin\beta}\right) - w_1 - w_2 = 0. \tag{15.4}$$

In this equation the only unknown is T_1 and thus a numerical value for T_1 is obtained. Once this is known a numerical value for T_2 can be determined using equation (15.1). With T_2 known a value for θ_e can be obtained since it is the only unknown in equation (15.3). Since the system is in equilibrium $|\theta_e - \theta_c| < \sigma_\theta$ should be true.

15.4 Procedure

The system shown in figure 15.1 is constructed. This image shows two poles that are supporting the system. These can be replaced by any two rigid objects that will support the weight of the experiment. The weight of a meter stick, w_1, and the hanging weight, w_2, are measured. The length l and the position of the weight r are also measured. It should be noted that l is the distance between the locations were the strings are tied to the stick and not necessarily the length of the stick.

The system shown in figure 15.1 is constructed such that $\alpha \neq \beta$ and $\theta > 0$. A camera is positioned perpendicular to this system and a photo is obtained. If the camera is off at an angle to the system then the measurements will be incorrect.

The image is then analyzed in a program such as GIMP. The angles α, β and θ are measured. In GIMP this tool is called 'Measure' and is turned on with Shift-M. Each of these angles will have uncertainties in their measurements.

15.5 Analysis

From the photograph the values $\alpha \pm \sigma_\alpha$, $\beta \pm \sigma_\beta$ and $\theta \pm \sigma_\theta$ are obtained. Furthermore, values $r \pm \sigma_r$, $l \pm \sigma_l$, $w_1 \pm \sigma_{w_1}$ and $w_2 \pm \sigma_{w_2}$ are measured directly.

Computing T_1 is accomplished with equation (15.4). Computing σ_{T_1}, however, can be very involved. Each uncertainty has a plus or minus value thus creating two choices for each. A quick and less accurate method of computing the uncertainty for σ_{T_1} is to find the combination of signs for each of the different values used in the equation that produce the maximum value of T_1. Then this process is repeated to find the minimum possible value for T_1. The uncertainty is the average of the differences $|T_1 - T_{1:\max}|$ and $|T_1 - T_{1:\min}|$ where T_1 is the value calculated in equation (15.4).

Once $T_1 \pm \sigma_{T_1}$ is known $T_2 \pm \sigma_{T_2}$ is calculated using equation (15.2) and standard error propagation. Since equation (15.3) is linear in $\cos \theta$, standard error propagation can also be used to compute $\theta_c \pm \sigma_{\theta_c}$.

The photograph will also yield the angle $\theta_e \pm \sigma_{\theta_e}$. The experiment is successful if θ_c and θ_e are equivalent within one standard deviation.

15.6 Possible problems

The possible problems for this experiment are:
- If θ is too small then it may be difficult to obtain a good reading in the software analysis.
- If the camera is not perpendicular to the experiment then the measurements of the angles will be incorrect.

Chapter 16

Lab 12: pendulum

16.1 Educational goal

The goal of this lab is to estimate the Earth's gravity using a pendulum.

16.2 Materials

The materials needed for this lab are:
- string,
- a table with a protruding lip,
- duct tape,
- a plastic bag (like a sandwich bag),
- a vertical pole (or similar as discussed in the text),
- a smart phone app that logs acceleration,
- (if possible) a smart phone app that logs both acceleration and orientation, and
- a spreadsheet.

16.3 Theory

The force equation for an oscillating pendulum without damping, shown in figure 16.1, can be written as

$$\frac{d^2\theta}{dt^2} + \frac{g}{L}\sin\theta = 0, \tag{16.1}$$

where θ is the angle of the pendulum from the vertical, g is gravity and L is the length of the pendulum. The pole in the figure will be discussed in section 16.4.

The first term is also angular acceleration $\alpha = \frac{d^2\theta}{dt^2}$. Furthermore, for an object moving about a circle of radius r the relationship between its linear acceleration,

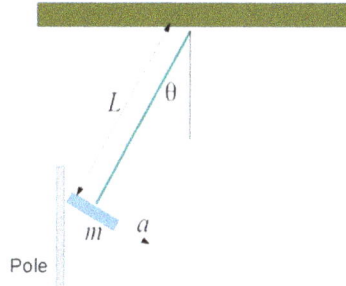

Figure 16.1. A simple pendulum.

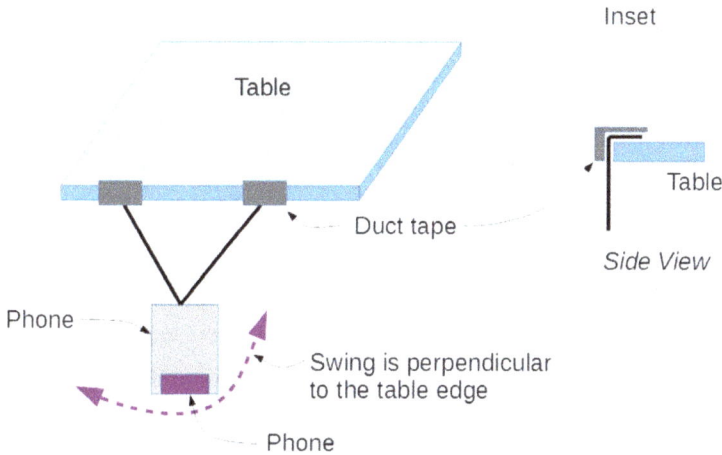

Figure 16.2. Attaching the pendulum to the table.

a, and its angular acceleration, α, is $a = \alpha r$. In this case, $a = \alpha L$. Thus, equation (16.1) reduces to

$$g = \frac{-a}{\sin \theta}. \tag{16.2}$$

It is interesting to note that neither the mass of the pendulum nor the length of the pendulum are included in this equation.

16.4 Procedure

The experimental system is shown in figure 16.2. The smart phone is placed in a plastic bag and the top of the bag is attached to two strings which are attached to the lip of the table. The reason that two strings are used is that with a single string the phone will tend to rotate about an axis that is parallel to the pendulum arm. The two string pendulum prevents this. The inset image shows that the string is attached to the table with duct tape and that this tape goes to the bottom of the table top edge. If the string were taped to just the top of the table then when it swings to the left the top of the pendulum will be at the top of the table and when it swings to the right then the top of the pendulum will be at the bottom of the table top. The phone is situated so that one of the axes is parallel to the arm of the pendulum and another axis is parallel to the motion of the pendulum.

There are two possible methods for performing the experiment depending on the phone app that is available. If the app has the ability to simultaneously record orientation and acceleration then the system construction is complete. The app is turned on and the pendulum is placed at some reasonable angle θ making sure that two of the phone's axes are parallel to the pendulum arm and the direction of motion. The pendulum is then released so that it can swing freely for several oscillations. The app is then turned off and the data are sent to a computer for analysis.

If the app only records acceleration data then a pole is set next to the pendulum as shown in figure 16.1. Its sole purpose is to ensure that the starting angle is known. Any rigid object can act as a pole. The orientation app is turned on and the pendulum is raised so that it just touches the pole, and this angle is recorded as θ. The orientation app is turned off and the acceleration logger is turned on. The phone is positioned such that one of its axes is parallel to the pendulum arm and another is parallel to the direction of motion. The phone is then brought to the pole and released allowing it to swing freely for a few oscillations. The logger is turned off and the acceleration data are sent to a computer for analysis.

16.5 Analysis

The analysis is basically to gather the information to compute equation (16.2). The raw data are shown in figure 16.3 which shows the oscillations. It is easily seen that there is a loss of energy as the pendulum swings as the peak heights become smaller. Also seen is that the first oscillation has some noise in it as the phone and bag settle into simple pendulum motion. Only one oscillation is required and it should be the first clean oscillation as delineated by the two areas. These data are isolated for further analysis.

The data shown here need to be shifted vertically so that the average of the oscillation values is 0. They also needs to be shifted horizontally so that the first data

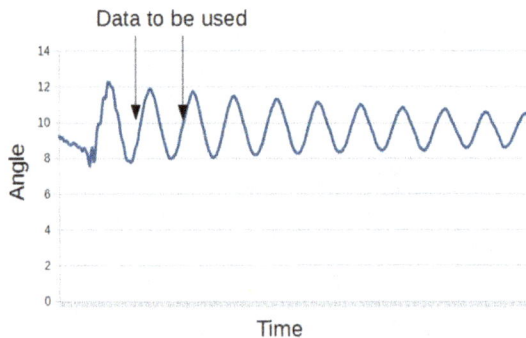

Figure 16.3. Graph of raw data acceleration along the direction of motion versus time.

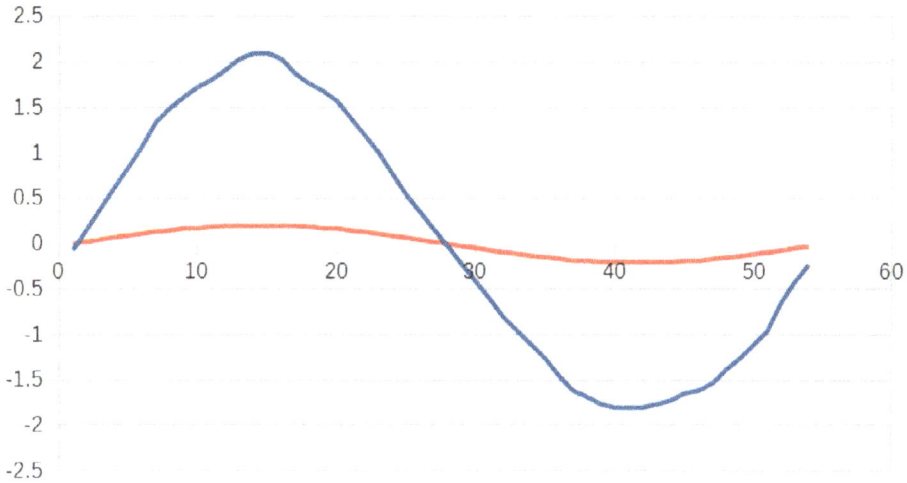

Figure 16.4. The portion of the data that would be used in analysis.

point collected is associated with time $t = 0$. This is shown in figure 16.4 which shows the acceleration data (larger graph) and the orientation data versus time. The numbers along the x-axis are sampling increments and in reality this oscillation occurred in just over 1 second.

If the app collected both the acceleration and the orientation then computation of equation (16.2) is straightforward. It should be noted that most orientation loggers store information in degrees while the spreadsheets perform calculations in radians. In the graph shown there are 53 data samples and equation (16.2) can be calculated for each time sample. The average and standard deviation of these values are then represented at $g_e \pm \sigma_{g_e}$ where the subscript indicates experimental.

If the app can only record the acceleration data then it is necessary to calculate the orientation angle. The starting angle is known and the amount of time that an oscillation requires is know from the acceleration data. The orientation data are then assumed to be sinusoidal and the intermediate values are calculated within the spreadsheet. With the simulated values of θ for each time increment the values for equation (16.2) are calculated. The average and standard deviation of these values are then represented as $g_e \pm \sigma_{g_e}$ where the subscript indicates experimental.

It is well known that the accepted value of gravity is 9.8 m s^{-2} and so the final step is to calculate the percent difference between the experimental value and the accepted value. A successful experiment would provide a value of g such that $|g - 9.8| < \sigma_g$.

16.6 Possible problems

These are some of the possible problems that could cause poor results.

- Smart phones record angles in degrees but the spreadsheet performs calculations in radians. The RADIANS() function in the spreadsheet will convert degrees to radians.

16-4

- The camera should be placed so that one axis is parallel to the pendulum arm and another is in line with the swinging motion. If the camera is tilted then the acceleration information will be spread amongst the different axes.
- The pivot point of the pendulum should be a single point. Incorrect attachment to the table top may create two pivot points which will be harmful to data collection.

Chapter 17

Lab 13: Kepler's third law

17.1 Educational goal

The goal of this lab is to study Kepler's third law by measuring the density of the Sun.

17.2 Materials

The materials needed for this lab are:
- a small hole in a piece of cardboard,
- a sunny day,
- a piece of poster board,
- a ruler,
- a meter stick, and
- a smart phone with an orientation app.

17.3 Theory

This theory follows the work of Mallman [1] in which there was a clever demonstration of measuring the density of the Sun using a tree. On a sunny day the Sun shines through the tree and some light is not blocked by the leaves but instead sneaks through between the leaves unabated. The result is a bunch of circles of light within the shadow of the tree on the ground. While some of the spots are brighter than others there is always a minimum diameter of these spots. This is due to the fact that the Sun is not a point source of light, but rather an extended one.

The leaves, or rather the gaps between them, act as a hole that the sunlight passes through. While this demonstration was enlightening, the use of a tree makes measurements difficult since the leaves move about in the wind and thus the spots on the ground move. So, for the lab the tree will be replaced by a single pinhole that is mounted at an elevated height which does not have to be measured.

doi:10.1088/978-1-6270-5628-1ch17
17-1

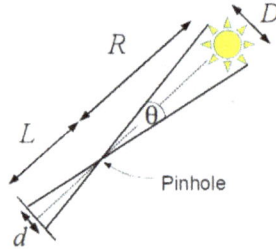

Figure 17.1. Sunlight passes through a hole making a spot on the ground.

The light shines through the hole as shown in figure 17.1. The diameter of the Sun is D, the distance from the Sun to the hole is R, the distance from the hole to the ground is L and the angle of the cone of light passing through the hole is θ. The diameter of the spot on a plane perpendicular to the direction of light propagation is d. As of now, none of these parameters are known and very few can be measured.

The goal is to measure the density of the Sun which will be assumed to be a homogeneous solid sphere. Thus, the density is

$$\rho = \frac{M}{V}, \tag{17.1}$$

where M is the mass of the Sun and V is the volume of the Sun. Since neither of these can be measured in this lab it will be necessary to define them in terms of variables that can be measured.

Kepler's third law can be written as

$$\frac{R^3}{T^2} = \frac{MG}{4\pi^2}, \tag{17.2}$$

where T is the orbital period (365.2 days) and G is the gravitational constant $6.67 \times 10^{-11}\,\text{N} \cdot \text{m}^2\,\text{kg}^{-2}$. This equation can be rewritten to solve for M and then substituted into equation (17.1).

The volume of a sphere is

$$V = \frac{\pi}{6}D^3, \tag{17.3}$$

where D is the diameter of the sphere. The relationship between a radius and an arc length is $D = R\theta$. These are also substituted into equation (17.1) to give

$$\rho = \frac{24\pi}{T^2 G \theta^3}. \tag{17.4}$$

The only variable that is not known is θ and this can be tricky to measure.

The geometry of the light after it passes through the hole is shown in figure 17.2. The height of the pinhole is H and the distance from the vertical to the spot on the

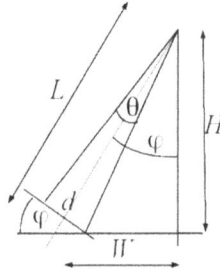

Figure 17.2. The geometry of the light after it passes through the hole.

ground is W. A screen is placed such that it is perpendicular to the incoming light and the diameter of the spot on the screen is d. The angle that this screen makes to the horizontal is ϕ.

By noting that $d = L\theta$ and $\sin \phi = \frac{W}{L}$ the density can now be written as

$$\rho = \frac{24\pi W^3}{T^2 G d^3 \sin^3 \phi},\tag{17.5}$$

where all terms on the right are either well known or easily measured.

The results can then be compared to the accepted value for the density of the Sun of 1400 kg m^{-3}.

17.4 Procedure

The procedure is fairly simple. A hole is cut into a piece of cardboard (or any dense material) and this cardboard is fixed at a height H which should be several meters. Since d is related to H better results are obtained with a larger H. It should be noted that it is not necessary to measure H but it is necessary to measure W. The distance, W, from the vertical location of the hole to the spot of Sun on the ground is measured with the meter stick (or tape measure).

The screen (poster board) is placed on the ground where the Sun shines through the hole. It is tilted so that it is perpendicular to the incoming light. This position should create a perfectly round spot of light rather than an elliptical spot. The angle of this board, ϕ, is measured with the orientation app on the smart phone.

17.5 Analysis

The analysis is simple. The measurement of ϕ will not be exact and therefore there will be an uncertainty σ_ϕ. There will also be an uncertainty σ_W in the measurement of the width.

These are used to compute $\rho \pm \sigma_\rho$ from equation (17.5) and proper error propagation. This is the experimental value and is to be compared to the accepted value of 1400 kg m^{-3}. The percent difference can then be calculated.

17.6 Possible problems

Possible problems with this lab include:
- The screen must be perpendicular to the incoming light or the value of d will be incorrect.

Bibliography

[1] Mallman A J 2013 Tree leaf shadows to the Sun's density: a surprising route *Phys. Teacher* **51** 10–11